# Unravelling the Credit Crunch

# CHAPMAN & HALL/CRC
Financial Mathematics Series

## Aims and scope:
The field of financial mathematics forms an ever-expanding slice of the financial sector. This series aims to capture new developments and summarize what is known over the whole spectrum of this field. It will include a broad range of textbooks, reference works and handbooks that are meant to appeal to both academics and practitioners. The inclusion of numerical code and concrete real-world examples is highly encouraged.

## Series Editors

M.A.H. Dempster
*Centre for Financial Research*
*Judge Business School*
*University of Cambridge*

Dilip B. Madan
*Robert H. Smith School of Business*
*University of Maryland*

Rama Cont
*Center for Financial Engineering*
*Columbia University*
*New York*

## Published Titles

American-Style Derivatives; Valuation and Computation, *Jerome Detemple*

Analysis, Geometry, and Modeling in Finance: Advanced Methods in Option Pricing, *Pierre Henry-Labordère*

Credit Risk: Models, Derivatives, and Management, *Niklas Wagner*

Engineering BGM, *Alan Brace*

Financial Modelling with Jump Processes, *Rama Cont and Peter Tankov*

Interest Rate Modeling: Theory and Practice, *Lixin Wu*

An Introduction to Credit Risk Modeling, *Christian Bluhm, Ludger Overbeck, and Christoph Wagner*

Introduction to Stochastic Calculus Applied to Finance, Second Edition, *Damien Lamberton and Bernard Lapeyre*

Numerical Methods for Finance, *John A. D. Appleby, David C. Edelman, and John J. H. Miller*

Portfolio Optimization and Performance Analysis, *Jean-Luc Prigent*

Quantitative Fund Management, *M. A. H. Dempster, Georg Pflug, and Gautam Mitra*

Robust Libor Modelling and Pricing of Derivative Products, *John Schoenmakers*

Structured Credit Portfolio Analysis, Baskets & CDOs, *Christian Bluhm and Ludger Overbeck*

Understanding Risk: The Theory and Practice of Financial Risk Management, *David Murphy*

Unravelling the Credit Crunch, *David Murphy*

Proposals for the series should be submitted to one of the series editors above or directly to:
**CRC Press, Taylor & Francis Group**
4th, Floor, Albert House
1-4 Singer Street
London EC2A 4BQ
UK

Chapman & Hall/CRC FINANCIAL MATHEMATICS SERIES

# Unravelling the Credit Crunch

## David Murphy

CRC Press
Taylor & Francis Group
Boca Raton   London   New York

CRC Press is an imprint of the
Taylor & Francis Group, an **informa** business

A CHAPMAN & HALL BOOK

Chapman & Hall/CRC
Taylor & Francis Group
6000 Broken Sound Parkway NW, Suite 300
Boca Raton, FL 33487-2742

© 2009 by Taylor and Francis Group, LLC
Chapman & Hall/CRC is an imprint of Taylor & Francis Group, an Informa business

No claim to original U.S. Government works

Printed in the United States of America on acid-free paper
10 9 8 7 6 5 4 3 2

International Standard Book Number: 978-1-4398-0258-8 (Paperback)

**Visit the Taylor & Francis Web site at**
**http://www.taylorandfrancis.com**

**and the CRC Press Web site at**
**http://www.crcpress.com**

# Contents

# List of Figures

# *Introduction*

## Playing in a Storm

The Credit Crunch is the most serious crisis in banking in several generations. Financial institutions have lost hundreds of billions of dollars, house prices have plummeted in countries from the U.S. to Asia, and millions of people will lose their homes. The prices of stocks and bonds have plunged. It is not just a matter of plummeting prices either: the Crunch has come off the financial pages and become a key social and political issue. It can even be argued that it was the deciding factor in the 2008 U.S. presidential election.

Given the turmoil, many people want something to be done. But before changing laws or implementing regulations – and before spending billions of dollars of taxpayers' money – it is important to understand why the problems happened. This book explains how the financial system was drawn into the Crunch and the issues that need to be addressed to prevent further disasters.

### Serious Games

My friend Steve is rather competitive. Ordinarily he is a nice person, but when he is playing a game – any game – he will do anything within the rules to win. Steve is smart, so he is usually adept at figuring out what move is best for him in any given situation. Some games are harder to play than others, but even when Steve is playing chess, he is pretty good at doing what is best for Steve. Of course, when Steve plays someone who is not as good at games as him, or someone who simply does not care about winning as much, he tends to beat them. That can be hurtful for the other party. Steve does not mean to upset them: he is simply doing his best within the rules of the game.

It is easy to attribute human qualities to financial institutions, to think that a bank 'wants to screw the little guy' or that a hedge fund 'is evil'. But often that is a mistake. Rather, they are participants in a large and complicated game: the game of finance. The way to understand what happens is to understand the rules of the game. Like Steve, financial institutions simply do what seems to be in their interest given the constraints they are under. Therefore

1

if we want to understand the Credit Crunch, we need to examine the rules that constrain how financial institutions operate: how they do business; what products are involved and how these products behave; how financial institutions raise money; and the legal and regulatory frameworks under which firms operate. This will occupy most of the book. Then, armed with an understanding of how the system works, we can discuss how to change the rules to make bad behaviours less likely.

## Reading this Book

This book is written for a non-specialist. I have tried to explain how the game was played, and how the rules of the financial system led to the Credit Crunch. Some of the features of the game are fairly esoteric, so they require a good deal of explanation, but I hope that none of them are literally incomprehensible. To this end there are no equations in the book, and all of the jargon in the main body of the text is explained as we go along.

Each chapter has end notes which give references and discuss some issues in greater detail, and sometimes at a greater level of sophistication. General readers will not miss the main arguments if these are skipped entirely.

## Timing

This book was written in the autumn and winter of 2008, up to January 2009 (and we discuss nothing beyond that date). In some cases, the book was written as the events described were happening. That gives little perspective on times that were, on occasion, tumultuous. Scholarship of the Great Depression did not really lay bare the underlying causes of that crisis until twenty or more years later, so there is certainly risk in discussing the near past. Nevertheless, our economic understanding is better than that of the 1930s. Systems thinking – ways of understanding the behaviour of complicated games like finance – has advanced too. So perhaps we can begin to understand how the actions of thousands of people like Steve working in financial institutions, regulatory bodies, central banks, and investment managers, led to the Crunch despite the lack of perspective.

## Acknowledgements

I should like to thank Miles Draycott for his continuing insights and perspective. Conversations with Jón Daníelsson, Anne McGeachin, Richard Metcalfe and Will Hall have been enjoyable and informative: for that, I thank them. I should also particularly like to thank Helmut Engelbrecht of Standard Bank and Patrick Fell and Matthias Sydow of the ECB for their invitations to present seminars at which some of these ideas were first discussed. Michael Dempster and my editor, Sunil Nair, have both been very supportive of this project: I am very grateful for their help and advice. Sara Jayne Farmer was a diligent and insightful proof reader and I thank her for that. All errors, omissions and mis-judgements remain, of course, my fault.

*Spitalfields*
January 2009

# Chapter 1

---

## What Happened?

*Here's the rule for bargains:*
*'Do other men, for they would do you.'*
*That's the true business precept.*
Charles Dickens

---

## Introduction

The Credit Crunch has displayed some dramatic phenomena. Banks have failed; markets have plunged; people have lost their homes. We begin by looking at what happened – and hence what we have to explain. Property is the obvious place to start, as falling house prices were one of the early symptoms of the Crunch and one of the major causes of losses. So we begin there. Then we look at the nature of the financial institutions involved in mortgage lending and how that has changed over the years. The actions of these firms transmitted the mortgage problem into a wider financial markets problem. Some parts of this market reaction are described, and we trace the linkages between mortgage borrowers, lenders, the financial markets and the wider economy[1].

The first stage of the Crunch involved U.S. residential house prices and those financial instruments that depended on them. This lasted from mid 2006 to mid 2007: during this period, it was clear that trouble was ahead, but the financial system was still functioning well.

The second stage began when the institutions that had mortgage risk suffered losses. This eroded confidence. Firms found it harder to raise money; they had less capital as it had been eroded by their losses; and they were worried about the credit quality of many of their borrowers. Hence they made fewer loans, and by early 2008 a general contraction of lending had begun. The Credit Crunch was visibly underway.

The third stage of the Crunch began in September 2008 with the failure of Lehman Brothers. Confidence had been fragile before this point: but when

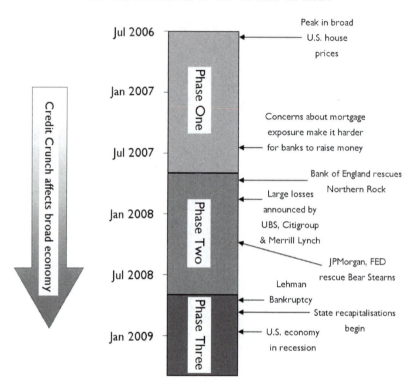

The Three Phases of the Credit Crunch

Figure 1.1: The development of the Credit Crunch

it became clear that the U.S. authorities would permit large institutions to fail, it was entirely lost.

There was something close to panic after the failure of Lehman. Governments now intervened, injecting new capital into banks, guaranteeing bank assets, and slashing rates. We survey the bailouts, and show how the taxpayer came to own much of the banking system.

## 1.1   U.S. Residential Property: The Crunch Begins

American housing plays a central rôle in the crunch. The Case-Shiller indices[2] provide aggregate information on American residential property prices for twenty different cities and on a national basis, so they are a good place to look if we want an overview of the American property market.

### 1.1.1  Average Prices

Figure 1.2 shows the twenty city composite index from 2000 to mid 2008. Average prices rose steadily during the first five years of the century and peaked in mid 2006. Therefore we put the start of the Crunch at this mid 2006 peak[3].

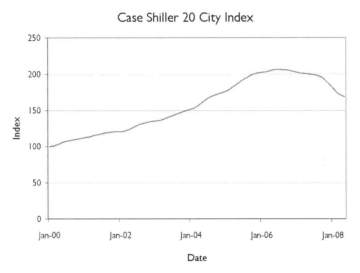

**Figure 1.2**: The S&P/Case-Shiller Composite Home Price Index 2000–mid 2008

By mid 2008 average house prices had fallen back to roughly 170% of their 2000 level. At first sight this does not seem too serious: a 70% return over eight years is not too bad, and while prices were down 20% from their high, they had not fallen calamitously by mid 2008. How can this relatively minor property correction have caused a banking crisis?

### 1.1.2  Worst Hit Areas

The problem becomes more obvious if we look at the individual cities. Figure 1.3 shows the data from three of the most badly affected metropolitan areas: Las Vegas, Los Angeles and Miami. Here, by 2006 prices had risen higher than the national average, but subsequently they have plunged much further.

If we look in more detail, the differences become even more extreme. Some of the suburbs and exurbs of Los Angeles, Las Vegas and Miami have seen property prices more than halve. By August 2008 `realtor.com` listed

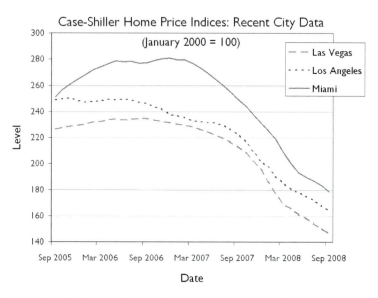

**Figure 1.3:** S&P/Case-Shiller Home Price Index data for selected cities, September 2005–2008

hundreds of family homes in these areas for less than $30,000[4]. If you are willing to live in the less desirable parts of Riverside, Dade or Clarke counties, then houses can be very cheap indeed.

### 1.1.3   *Speculate to Accumulate*

Why did these areas – and some others, mostly in the mid-West – suffer much more than the Eastern seaboard?

The answers are trend and speculation. The broad population trend in the U.S. is towards the sun. For instance, the three fastest growing states from 2000–2005 were Nevada, Arizona and Florida[5]. Clearly new housing is needed where population is growing. Moreover, the strong growth in house building typically happens where there is significant demand and property prices are rising fast: it is here that the biggest profits are available for house builders. Providing demand continues, a spiral can be created:

- rising prices encourage speculative development;

- this tends to happen on the margin;

- infrastructure develops to serve these marginal developments, making them more attractive;

- thus prices continue to rise, and new developments begin even further out.

This continues until the oversupply of property is so large that the market is glutted.

Builders are not the only speculators: seeing a rising house price trend, some individuals will buy new or even partially built properties hoping to flip them onto the market for a profit a few months or years later. This too encourages prices to rise.

### 1.1.4   Leverage

Speculation would not be too dangerous if people used their own money. Property prices cannot go up forever: at some point they would start to fall and speculators would lose money. This would be painful for them. Moreover, job losses at builders, realtors, mortgage brokers and so on would be bad for the local economy. Typically, though, the effects would be limited. Sadly, though, most property speculation involves *leverage*.

Leverage is used whenever an asset is bought with borrowed money. The most obvious example is a mortgage: money is borrowed to buy a house, and that money has to be repaid over the term of the mortgage. If it isn't, the lender can seize the property and sell it to recover the amount they are owed. For the moment let us assume that this is all the lender can do – they have no *recourse* to the borrower's other assets.

The problem with leverage is that it allows failed speculation to affect more than the speculator. Suppose that I had bought a new family house in the suburbs of Fort Myers, Florida for $250,000 (a typical amount in early 2006), and that I taken out a non-recourse mortgage of $225,000. By mid 2008, the house is worth $160,000. If I default on my mortgage, the lender will lose at least $65,000 (the difference between the mortgage amount and the value of the house). If things had gone well, I would have a profit. But when there are losses, I am not the only one who can lose.

Leverage creates an incentive to speculate. If property prices had gone up 36%, I would have made $90,000 on a stake of $25,000: but when they fall by the same amount, I only lose my $25,000 stake. This 'heads I win a lot, tails I lose a little' feature means that when prices are rising, speculators are drawn into the market, creating a price bubble. When this bursts, it is not just the speculators who lose money but also the lenders who helped them to finance their properties.

### 1.1.5  *Mortgage Foreclosures*

Suppose a borrower fails to pay a scheduled mortgage payment for a short period of time: a few weeks, say. The mortgage is then said to be *delinquent*. The missed amount plus interest is added to the amount owed. A fee might also be imposed. As the period of non-payment lengthens, the lender will begin to take action, first sending a letter or otherwise contacting the borrower, then declaring a formal breach of the terms of the mortgage. Finally, if non-payment persists and nothing more can be done, the lender *forecloses* on the mortgage.

Foreclosure is a process which allows the lender to seize the property: depending on the jurisdiction[6] it might require a court order (a process known as judicial foreclosure) or simply that the borrower is notified that they are in default. The borrower typically then has an extended period of time to *cure* the default – to make good on the payments – or to try to agree a *modification* of the loan.

Foreclosure is the end of the road. A lender can typically only foreclose after the borrower has been given some time to pay, and even then they will only seize the property if there is no reasonable alternative such as changing the terms of the loan. Foreclosure rates therefore give an indication of completely failed mortgages. Since 2006 these have rocketed in the worst affected states. According to RealtyTrac, for instance[7], one in every 43 Nevada households received a foreclosure filing during the second quarter of 2008, up from one in 452 in September 2006. These rates are historically unprecedented: they reflect not just individual disasters, but also whole areas of the United States which now stand empty.

Foreclosure only generates a loss if the property is sold for less than the mortgage balance. Falling house prices therefore lead to much higher losses to mortgage banks: as figure 1.4 indicates, by 2008 these loss rates (sometimes called *charge-off* rates) were at the highest level for many years[8].

Many states give lenders some freedom in deciding when to sell a foreclosed property. In California for instance a year's delay is permitted. However if prices are falling, lenders will want to try to sell quickly. This is especially true as many lenders only have limited resources to manage *real estate owned* (REO). With rising foreclosure volumes they tend to sell what they can to keep the amount of REO from rising too fast.

In a foreclosure the lender has the right to seize the house. What they do not have, typically, is the right to seize what is in the house. Therefore the borrower will often strip the place to maximise their return, taking doors,

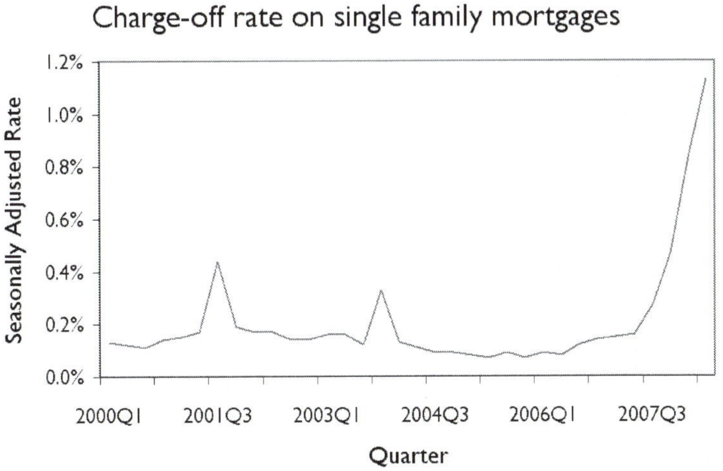

## Charge-off rate on single family mortgages

Figure 1.4: The charge-off rate for residential mortgages for U.S. commercial banks

light fittings, even sometimes baths and basins. This means that foreclosed properties are often not in a saleable condition, or only saleable at a large discount. Empty properties – especially empty properties without doors – are susceptible to vandalism.

Both of these phenomena mean that foreclosure depresses house prices. Lenders would rather sell fast if they can and take a small loss than hang on and get a better return several years down the line, especially given that they have to pay property taxes and insurance in the meantime. Property that cannot be sold blights the neighbourhood, reducing the value of adjacent properties. This in turn encourages further mortgage defaults.

The system, then, has built in accelerators. On the way up rising prices encourage speculation, driving prices up further. On the way down foreclosures rise, driving prices down. Home price stability does not naturally arise as a result of this particular game.

### 1.1.6 Property-related Activity

Declining property prices and rising foreclosures create two problems. First, any company whose income depends on property prices makes less money. Thus builders can sell new houses for less, realtors take less commission, and mortgage fees (which are typically a percentage of the amount lent) are lower. Second, foreclosure produces an overhang of inventory. This competes with newly built houses and house price sales by private owners.

Speculative construction becomes unattractive: not only is the price a house is likely to sell for lower in the future, but it will also be harder to sell. Therefore as the Crunch bit, U.S. house building contracted dramatically, with new home starts falling to a 17 year low by July 2008[9]. Homebuilders' confidence in the housing market fell to record lows in the same period[10].

The most important impact of residential property on the broader economy is not its price: it is the volume of house sales[11]. The reason is that even relatively small declines in house prices tend to be associated with plummeting volumes. Home owners are reluctant to take a loss on sale so they don't sell, and any businesses which rely on volume – like realtors or mortgage brokers – make much less money. Some people have to move so they rent out their home, resulting in falling rental rates as supply exceeds demand. This in turn pressures buy-to-let landlords whose income may now be insufficient to pay for their mortgages. Even worse, home owners see that they have less equity in their properties and hence their sense of prosperity declines. They consume less and this affects many sectors of the economy.

Once volumes and prices are falling, the effects are persistent. Few people want to buy a property that they think will be worth substantially less in a year, so buyers withdraw from the market, particularly first time buyers. Income growth slows as the economy flirts with recession, again reducing confidence. Major purchasing decisions are put on hold. House prices will not recover until people believe that the economy is re-established on an upward trend.

### 1.1.7  Residential Property Outside the United States

The United States residential property market is not unique in suffering from a bursting bubble. The residential property markets of several other countries have had problems. None of these has caused anything like the problems of the U.S. residential downturn, but it is worth looking at a couple of representative examples: Ireland and Spain.

The two countries had similar experiences in many ways. Both started the 1990s as relatively unsophisticated economies with fairly high interest rates: both grew fast after entry into the Eurozone, partly stimulated by much lower interest rates. This growth caused property prices to soar: both Spain and Ireland's property markets more than doubled since Eurozone accession. It also attracted migrant workers, putting further pressure on housing. Rising prices encouraged speculative building, and house construction became an important sector of both economies. In Ireland in 2007, for instance, the construction sector represented 13% of jobs and was connected to roughly a quarter of GDP[12].

One effect of all of this construction was to boost home ownership to unprecedented levels: 77% in Ireland and over 80% in Spain. Another, as prices rose, was to increase household indebtedness. Both countries' economies had become over-exposed to rising property prices, and both relied on low interest rates to sustain the boom.

This could not last. The European Central Bank raised rates from 2.25% in late 2005 to 4.25% in mid 2008. In both countries most mortgages were variable rate, so mortgage costs went up. The effects of higher costs quickly appeared:

- Delinquencies rose fast. By early 2008, doubtful mortgage assets in the Spanish banking system were growing at the fastest rate since 1993[13].

- House prices fell, and house construction slowed dramatically. Buyers disappeared. This caused a glut of unsold houses: by September 2008, estimates were that one and a half million homes were unsold in Spain.

- The construction downturn resulted in increased unemployment as the building trade laid off workers: both Irish and Spanish unemployment hit ten year highs by late 2008. GDP growth slowed too, with both economies flirting with recession.

- Over-extended construction firms began to fail. A prominent example was the demise in July 2008 of one of Spain's largest property developers, Martinsa-Fadesa. It failed to persuade investors to provide it with more financing and was forced to declare bankruptcy.

The economic effects of the property price crash will be felt in both economies for some years. However, neither were as badly affected as the United States: the Irish and Spanish examples are mini-Crunches rather than full blown disasters.

## 1.2 Old and New Style Banking

If I need to borrow some money for a week and you have extra cash in your current account, it is very unlikely that we will agree a transaction. How will we find each other? Can you trust me to repay you? These problems are solved by the use of a skilled and trusted intermediary: a bank.

The bank pays you interest for the funds in your account and in exchange takes your money and lends it to someone else: in this case, me. Don't be fooled by that comforting positive balance on your statement: the money has long since gone somewhere else. I pay interest to the bank for my loan: typically they get more from me than they pay to you, and so the bank makes money. Financial institutions therefore play a key rôle as intermediaries in the financial system. If we want to understand how the system works, and how it can fail, we need to understand their strategies.

### 1.2.1 Banking as Borrowing Short and Lending Long

There is an old-fashioned rule which tells bankers how to be successful. The short form of it is:

$$3 - 6 - 3$$

This is shorthand for the following strategy:

- Borrow at 3%;

- Lend at 6%;

- Be on the golf course at 3pm.

In other words, to be successful as a banker, you need first to borrow money. Without *funding* you cannot do anything. The easiest way to get money is to take deposits. Then once you have the cash, you lend it out at a higher rate. Provided enough of the people you lend to pay you back, you will make money. This is a pretty simple strategy, so executing it does not trouble bankers too much, and they can begin 'client entertaining' at 3pm.

This strategy was how many banks worked until the late 1990s, so we call it *old style* banking. Notice that it involves making loans and keeping them: once the loan has been made, the bank keeps the risk on its balance sheet until either the loan is repaid or it defaults.

### 1.2.2 How Safe is an Old Style Bank?

Old style banks are vulnerable to credit risk. They have lent people money and if those people do not pay them back, they suffer. In particular, if they have lent people money to buy houses, and those people do not pay their mortgages, then they lose money if they cannot sell the house in a foreclosure for more than the amount lent. Rising house prices makes mortgage banking rather low risk. A falling house price environment is much worse: by the time the bank has gone through the foreclosure process, the house may be worth much less than the mortgage amount.

Suppose a bank takes £100,000,000 of deposits and makes loans for the same amount. How safe is the bank? It has an obligation to repay the depositor whenever they ask for their money. But it has lent their money to someone else. If any of the loans turn out to be bad, that might be a problem. In order to keep banks reasonably safe, they are required to have sufficient *capital*. This is money that does not have to be paid back: typically it comes from issuing stock and from retaining earnings. Thus our bank with £100,000,000 of loans might have a *capital requirement* set by its regulators of £8,000,000. This capital allows it to *absorb losses*, so providing no more than 8% of the loans default, the bank will be solvent[14].

### 1.2.3   The Originate and Distribute Model

One of the problems with old style banks concerns the capital they are required to have. Bank shareholders want a good return on the capital they have contributed. But the requirement for the bank to keep enough capital against its loans means that that return of capital cannot be very high. To see this, suppose the bank followed the 3–6–3 rule. Then the gross income from the loan book would be 3% of £100,000,000 or £3,000,000. The bank will want to keep some reserve against non performing loans as it would not be prudent to assume all of the loans will pay. Suppose this *loan loss reserve* is 1% of the total or £1,000,000. That leaves a net income of £2,000,000.

What is the bank's return on equity? The return is £2,000,000, but regulators demand at least £8,000,000 of equity, so the ROE is 25% before any taxes have been paid, or less than twenty percent after tax. This is not very impressive. If the lending is prudent, then the bank makes money, but this kind of banking is a fairly pedestrian, low return business thanks to the capital required to support the loan book.

Until the mid 90s there was very little banks could do about this. They engaged in four activities which where inextricably interlinked:

- *Originating* loans, that is finding people who wanted to borrow money;
- *Funding* loans, that is raising money to make the loan;
- *Servicing* the loans, that is collecting interest payments, chasing up late payments, modifying the terms of the loans to maximise returns and so on; and
- *Taking the risk* of loan default.

It was only the last of these that gave rise to capital requirements, so an obvious way to increase the bank's ROE would be to pass on the risk of the loan to someone else. But there was a problem: most of the loans banks make are to parties whose credit risk is not well known. If they had lent to Ford Motor Company, say, then it would be easy to sell the loan and so get rid of the troublesome capital requirement, because many buyers are in a reasonable position to assess the credit of Ford. But if they had lent money to me, any buyer would want to know how good a credit I was – how likely I am to pay the loan back – and it isn't worth finding that out for a measly little £1,000 loan, or even a £200,000 mortgage.

The solution is to sell not single loans, but a portfolio of loans. Then the details of individual borrowers matter less and the risk of the portfolio can perhaps be described by some summary statistics. For instance the bank might have a billion pound collection of UK mortgages, none individually over half a million pounds, all made to people who were considered good credits, none with any delinquencies in the last year. Perhaps a portfolio like that might be saleable?

At first the potential buyers were not convinced. How could they be sure the bank was not selling them a bad portfolio – just those loans that the bank was worried about for some reason? The issue is *alignment of interests*. The bank's interest is best served by selling loans that will turn out bad. The buyer's interest is best served by buying loans that will turn out good. The buyer is suspicious of the seller and finds it difficult to get enough information to reassure themselves. The seller does have a good reason to sell the loans – freeing up capital – but also a bad one – getting rid of bad loans. How can the buyer figure out what is a good transaction?

The solution to the problem is to give the buyer some additional comfort that the seller has sound motives. Suppose the expected losses on the portfolio are one percent as before, based on the historical performance of similar kinds of loans. If the bank agrees that it will keep the first 4% of losses, then it still has considerable exposure to the portfolio, and the seller is reassured. The bank has an interest in keeping losses on the portfolio low since otherwise it will suffer. In particular if the bank continues to service the portfolio, then the buyer of the risk has a reason to believe that this will be done diligently, since the selling bank bears the consequences of bad servicing.

The idea of banks selling a portfolio of loans but keeping the risk of some losses led to a change in the character of banking. Because the bank no longer had all of the risk of the portfolio, some capital was freed up. This allowed banks to increase their ROEs. *New style banking* began with the strategy:

- *Originate* loans;
- *Service* the loans (and take a fee for doing so);
- *Take the risk* of some loan defaults;
- But *Distribute* some of it to third parties.

This strategy is sometimes known as the *Originate to Distribute Model* or ODM. Pure ODM banks are in the business of moving risk rather than keeping it. Their core competence narrows to originating loans and servicing them, with only enough risk management being needed to deal with the retained risks. This risk retention, at least in the pure ODM, is reduced as much as possible.

The ODM also generates business for intermediaries. Various parcels of risk are traded, split and repackaged. The intermediaries assist in this process. They are sometimes said to *provide liquidity* to the market in that they will sometimes buy parcels in the hope of being able to sell them on at a profit. This means that rather than risk being held to term by the originator – as it is in an old style bank – risks may be traded a number of times.

The active trading of risk means that it is hard to know what an institution's exposures really are. It is relatively easy to understand an old style bank because all of its risk is on its balance sheet. A few summary statistics of loan types, maturities, the credit quality of counterparties, and so on will give a reasonably good picture. But with an ODM bank which has retained some risks but sold others, and which may have bought risks originated by others, the institution's exposure is more complex and more difficult to discern.

### 1.2.4 *Ratings and Lemons*

The *Lemons problem* is well known to second hand car buyers[15]. The issue is that some second hand cars are lemons and some are not. But finding out which ones are good is difficult and expensive. So buyers bid down the prices of all second hand cars because they have no easy way of identifying lemons. This is an example of a problem of *asymmetric information*: the seller knows how good the car is, but the buyer has no reason to believe what the seller tells them.

We solved the asymmetric information problem for loan portfolios in the previous section by requiring the seller to take a first loss position. In second hand car terms this would be like the seller pledging to pay the first £2,000 of maintenance bills on the car for two years. It does not remove the buyer's risk, but it does mean that the seller suffers first if there is a problem.

Another way of dealing with asymmetric information is if there is a third party who can be trusted to perform a disinterested analysis of the risk. In second hand car terms this would be some organisation – perhaps the manufacturer of the car – who puts a seal of approval on second hand cars. Provided the third party is both independent and knowledgeable, the buyer can get some comfort from their analysis.

The *Ratings Agencies* attempt to play this rôle for packages of loan risk. They perform a review of types of loans in the portfolio and how the risks of the portfolio are passed on to the buyers. Based on this analysis, they provide an assessment of the quality of the risk a buyer would take. This assessment is presented on a scale from AAA (the lowest risk) through AA, A, and BBB (all good quality or *investment grade* risk; then down to BB, B, and CCC.

### 1.2.5    *Playing ODM: When Interests are Not Aligned*

The ODM is one example of a trend towards specialisation that has been one of the key features of financial services for the last quarter of a century. Certainly the attributes that a firm needs to make loans – a well known brand and a sales force to attract business – are rather different from those it needs to keep the risk of those loans. Risk is best retained by those able to understand it, and with the appetite to take it. In the case of the long term credit risk of a book of loans, that might well be a pension fund or some other form of investment manager. So while the ODM grew up because banks were trying to improve their ROEs, it is not necessarily bad for the financial system. It did at least encourage risks to move to a natural home.

New style banking becomes dangerous if interests are not aligned. Let us suppose that the size of the risk retained by a bank falls over time. One year it is 4%, the next 3.5% and so on until eventually buyers of loan portfolios are not demanding any meaningful risk retention at all. This might happen because times are good and the loans are performing well. In those circumstances buyers are willing to take more risk as the dangers seem minimal based on recent history.

Once alignment of interests disappears, the game changes dramatically. Now the banks do not need to originate good loans for them to be saleable. They just need to originate *any* loans. Moreover, since their profitability is based on the amount of loans they originate and service, they are incentivised to keep finding people to borrow money. If loan portfolio buyers keep buying despite falling retention and declining portfolio quality, then the banks will carry on making money. This is one of the reasons for the Crunch: as alignment of interests failed, loans were made which would never have been contemplated by an old style bank.

## 1.3    What Happened in the Markets: The Second Stage

The capital markets are where institutions trade packages of risk. Sometimes this risk is in the form of securities, such as bonds or stocks; on other occasions the risk is transferred by other means such as derivatives or insurance.

One of the features of the Crunch has been a considerable dislocation in the capital markets: the prices of many assets have moved significantly. Moreover, assets which used to be easily tradable (and so had a readily observable price) became illiquid as market participants shunned them. In this section we look at the broad trends in some parts of the capital markets: these are some of the phenomena that will have to be explained if we want to understand the Crunch.

### 1.3.1  Trends in the ABX

One of the features of the ODM is that, since so many banks packaged up and sold the loans that they had made, there are readily observable prices for some portfolios. In particular the *ABX indices* represent the prices of portfolios of low-quality U.S. mortgage loans, so they give us an insight into the price at which market participants are willing to trade this kind of risk[16].

The ABX indices come with two characteristics. One is the six month period during which the underlying mortgages were originated. Thus the 06-02 index represents the prices of packages of mortgages originated during the second half of 2006. The second dimension is the credit rating of the risk, so the ABX AAA 06-02 is an index of packages of mortgages rated AAA when they were first sold.

Figure 1.5 shows the price history of the ABX AAA 0602s and 0702s. It raises an obvious question. How did AAA bonds – supposedly the safest type of security – lose half their value in less than a year?

There are three parts to the answer:

- As property prices fell, mortgage delinquencies rose, and hence the risk of loss on portfolios of mortgages increased significantly;

- Uncertainty increased too. No one knew how many of the mortgages would ultimately default, and hence the long term value of packages of mortgage risk was unclear. Investors hate uncertainty, and so they punished the sellers of these packages by assuming a plausible worst case;

- Many market participants withdrew from the market because of this uncertainty. Liquidity plunged: it became very difficult to find buyers of mortgage risk. This in turn depressed prices as investors demanded compensation for bearing this liquidity risk.

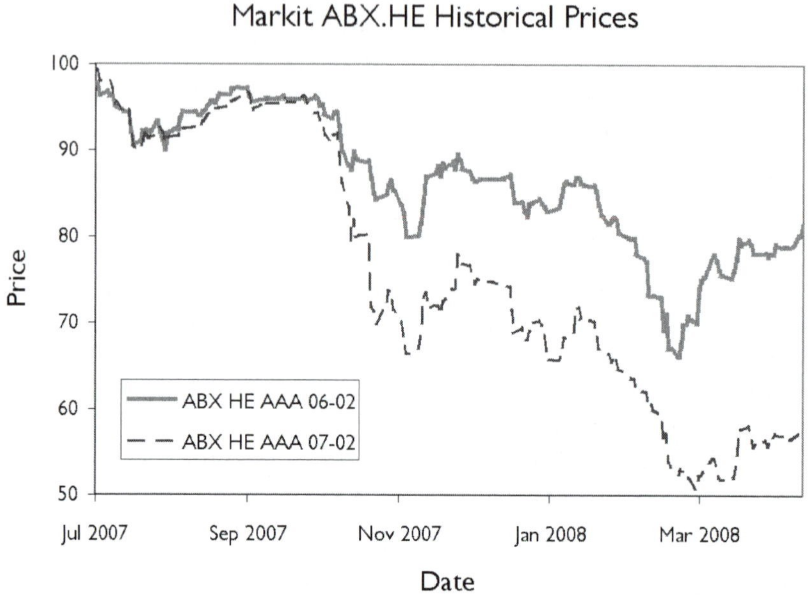

## Markit ABX.HE Historical Prices

**Figure 1.5**: Price history of two of the Markit ABX AAA Indices during the first and second phases of the Crunch

### 1.3.2   Short Rate

The availability of credit is vital to the economy. People buy houses and goods on credit; companies borrow to invest in research and manufacturing capacity; governments build infrastructure using borrowed money. Therefore when it is harder to get credit – where there is less of it and borrowing is more expensive – the effects on the economy can be considerable. Growth slows and unemployment rises.

Credit is supplied by the financial system, and so a healthy financial system is vital to a well-functioning economy. The Crunch started as a U.S. mortgage problem. Since much of the risk of these mortgages was held by financial institutions around the world, their health was affected. These firms suffered losses as the prices of packages of mortgages fell. Because no one knew how big institutions' losses would turn out to be, firms became reluctant to lend to each other. The credit quality of many of the world's financial institutions fell. Some of these institutions started to hoard cash rather than lend it out. This in turn meant that the real economy was starved of credit.

There are two kinds of interest rate which are important in determining the cost of borrowing: government rates and interbank rates. Government rates are determined by the compensation market participants demand for lending to the government. In particular in all major currencies there is a key government rate known as the *short rate*. This is the rate at which the Government lends money to the financial system for a short period. It is set to manage inflation: when inflation is low and stable, the short rate can be fairly low; but if inflation starts to spike up and the economy is growing too fast, borrowing is made more expensive by increasing the short rate.

This process is usually managed by a central bank. The precise criteria vary from country to country. For instance, the U.S. Federal Reserve (FED) simply has a mandate to ensure 'price stability', while the European Central Bank (ECB) targets inflation in the Eurozone of 2%. In both cases the short rate is set to meet the desired economic objective.

In ordinary conditions, lowering the short rate stimulates the economy. This is because lower short rates make it cheaper for banks to borrow; they pass these savings on to their clients; thus credit becomes cheaper in the economy as a whole and growth is stimulated.

Some central banks reacted to the Crunch by lowering short rate. For instance figure 1.6 shows the short rate in U.S. dollars known as FED funds[17]. The FED attempted to mitigate the effects of the failing property market by making it cheaper for banks to borrow money[18].

**Figure 1.6**: The short rate and three month Libor in U.S. Dollars

### 1.3.3  Libor

Cutting the short rate did not work very well. This is because most of the money banks need – their *funding* – comes from other financial institutions rather than the central bank. One measure of the rate at which money is lent between institutions is the London Inter-Bank Offered Rate or Libor. Libor rates are quoted for borrowing for various lengths of time, and the three month Libor rate is a particularly important benchmark.

In ordinary conditions the difference between the short rate and three month Libor is small and stable[19]. If the central bank cuts the short rate, banks borrow from the central bank and lend to the broader financial system, bringing Libor down too. However in the Crunch, banks became reluctant to lend to the wider financial system. This was partly because there was much more uncertainty in their own funding needs, and partly because they did not know what the real exposure of institutions really was thanks to the pass-the-parcel nature of the ODM. Lacking good information, they were reluctant lenders. This meant that Libor rates remained stubbornly high despite falling short rates, and so the cost of borrowing did not decrease. Moreover, as figure 1.6 illustrates, Libor was rather volatile. The interbank system became less efficient, and borrowing became more difficult, more uncertain, and more costly.

### 1.3.4  The Wider Debt Markets

The same firms were important both as intermediaries in the market for packages of loans and in the broader bond markets. The key players were either American broker/dealers such as Goldman Sachs, Lehman Brothers and Merrill Lynch, or large banks such as Citigroup, JP Morgan, Deutsche Bank and UBS[20]. Things did not use to be this way: twenty years ago many of the important players in French Franc bonds, for instance, were French. But now the same global firms provide liquidity to many different markets.

This might be good for clients – they do not need to call a French firm to buy a French bond – but it does mean that when these firms withdraw from the market, liquidity drops everywhere. This is what happened in the Crunch. Many of the global debt markets traders suffered losses due to their exposure to packages of U.S. mortgages. This made them more risk adverse across all of their business, and they reduced their risk taking. Faced with less liquidity, more uncertainty, and a worsening economic environment, bond buyers began to demand more return. Thus a range of different bond markets were affected:

- Packages of *commercial property* risk were affected as investors reduced exposure to all types of property.

- Similarly confidence in all types of instruments issued by ODM banks fell, impacting the prices of packages of credit card and other unsecured retail lending risk, auto loan risk, and commercial loan risk.

- As it became clear that the Crunch was not just a problem for banks and U.S. mortgage holders, but also for the wider economy in the United States and elsewhere, investors started to demand more interest from corporates who wanted to borrow money. Credit costs increased across the spectrum, with the effect being particularly noticeable for lower quality borrowers.

### 1.3.5 Stocks

Many bonds fell in value during the second phase of the Credit Crunch and liquidity in the bond markets has been seriously affected. The equity markets, by contrast, did not see calamitous falls at first. Consider for instance figure 1.7 showing the S&P 500 index from 2004 to mid 2008. The market went up steadily until the impact of the Crunch became obvious. Subsequently there has been increased volatility and some market falls, but the broad equity market held up reasonably until the third phase.

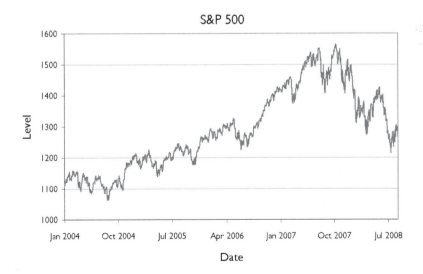

**Figure 1.7:** The S&P 500 Index 2004–mid 2008

The differing reactions of corporate bonds and stocks during the second phase of the Crunch is paradoxical to a believer in efficient markets. According to the bond market, we were going to experience a long and deep recession with a significant number of corporate bankruptcies. According to the equity market, things were not quite as good as they were, but there was no cause for panic.

Within that broad equity market optimism, there were significant winners and losers. Those financial and other institutions which were particularly exposed to U.S. residential property suffered significantly. In particular, two prominent institutions failed during this second period of the Crunch.

Northern Rock was the first to go. This was a bank which had embraced the ODM. It had expanded its mortgage business fast. When the crisis hit, Northern Rock was seen to be over-extended: other banks would not lend it money, and the Bank of England was forced to intervene. After a period of Central Bank support, Northern Rock was eventually fully nationalised in February 2008.

Bear Stearns was the second large casualty. There were five large broker/dealers in the U.S., and Bear Stearns, known colloquially as the Bear, was the smallest of these. One of the Bear's specialities was packages of mortgage risk: if you were in the market for this kind of bond, Bear Stearns would have been a natural firm to call. Therefore it is hardly surprising that the market lost confidence in the Bear as the market for these instruments fell and liquidity dried up. The FED became concerned that a Bear bankruptcy would result in the firm's huge inventory of mortgage risk being dumped into an already-depressed market. Another issue may have been the Bear's heavy involvement with hedge funds: had it failed, its trading partners would have suffered losses, and some of them would have failed too. Given these connections, the FED arranged a takeover, and JPMorgan bought Bear Stearns in March 2008.

## 1.3.6    Special Purpose Failures

Northern Rock and Bear Stearns were not the only entities to disappear during the second phase of the Crunch. There were a number of other company failures[21], including several *conduits* and *SIVs*. These were special purpose companies set up to enhance banks' earnings. The basic idea was to move assets from a bank's balance sheet into a new vehicle. The vehicle would pay for most of the assets by issuing debt to third parties. Under certain conditions[22], the bank appears to have sold the assets and hence it looks less risky to investors and to regulators. Nevertheless the bank can retain some benefit from the transferred assets by taking a stake in the special purpose company: it might, for instance, own some of its equity.

The value of the assets in many of these vehicles fell. Some SIVs and conduits could not repay their debt, and so they defaulted, giving losses for their investors. However in other cases – especially where the vehicle had been sponsored by a large bank – there was a different situation. Often the bank had undertaken to provide liquidity to the vehicle, that is to lend it money if no one wanted to buy the vehicle's debts. These liquidity lines resulted in SIV and conduit assets coming back on banks' balance sheets.

Furthermore, even where the buyers of the vehicle's debt were legally liable for losses, they pointed out to the sponsoring bank that the debt had often been sold as a low risk investment, and that if the bank pressed its case then they would be unlikely to find favour in the bond markets again. Given that the banks needed the bond markets more than the bond markets needed the banks, many sponsors capitulated and provided support to their vehicles so that the bond investors did not lose money. This support was a gift to the bond markets, offered in the hope of later favour. It was, of course, in stark contrast to the paucity of help available to suffering mortgage holders.

## 1.4 Après Lehman le Déluge: The Third Stage

Property prices in the U.S. continued to fall during 2008. Mortgage delinquencies rose, not just on bad quality loans, but also on higher quality borrowing: everyone was subprime now. Mortgage losses damaged two government sponsored entities active in the mortgage market, *Fannie Mae* and *Freddie Mac*. The U.S. government was forced to step in to save these entities.

The Crunch now gathered pace: in rapid succession all four of the remaining broker/dealers came under fire; a huge insurance company, AIG, was forced to ask the FED for an emergency loan; and the equity markets fell dramatically. Piecemeal government bailouts were no longer enough at that point. Trillions of dollars of assistance were needed, and most large banks received some form of state aid. The socialisation of finance had begun.

### 1.4.1 *Freddie and Fannie Fail*

Fannie Mae and Freddie Mac are peculiarly American hybrids: they were owned by private shareholders, despite being sponsored by the government and chartered by an Act of Congress. Before the Crunch began, their debt was often thought to be equivalent in quality to the U.S. government's. They were

set up to support the American housing market, and they are enormous holders of mortgage risk: between them, they held the risk of approximately five
trillion dollars worth of mortgages. Because they are not banks, Fannie and
Freddie do not have to keep as much capital for their risks as a bank would.
This allowed them to be massively leveraged. When mortgage losses started to
rise, Freddie and Fannie lost money, threatening their solvency. Investor confidence had been wrecked by earlier failures, and the markets began to doubt
that Freddie and Fannie really were government backed. The price of their
debt fell and their ability to raise money was seriously affected. Given that
Fannie and Freddie were vital to the U.S. mortgage market, the government
had to intervene. On 7th September 2008, Freddie and Fannie were placed
into a form of rehabilitation known as conservatorship, both companies being run by their regulator. The U.S. Treasury also provided various forms
of assistance to the two entities, lending them money, injecting new capital,
and undertaking to buy their debt if no one else would. The U.S. public debt
increased by 50% at a stroke[23].

### 1.4.2   Lehman Loses

After the fall of Fannie and Freddie, anyone with high leverage and lots of
real estate exposure was suspect. Lehman Brothers was an obvious candidate.
It was a broker/dealer, like the Bear. Also like the Bear, it had significant exposure to mortgages. Lehman took both significant residential and commercial
mortgage risk, so clearly 2008 was not a good year for it.

The firm posted significant losses during the second quarter of 2008. It
needed more capital to restore confidence. Various parties were reported to be
interested in buying stakes in it, and the firm was also rumoured to be raising
cash by selling its asset management arm. Neither move was consummated,
Lehman's share price tumbled, and the firm started to have difficulty funding
itself.

Firms which rely on confidence sensitive funding and then lose the confidence of the market fail quickly. On the 9th September 2008, Lehman's shares
fell by 45%: see figure 1.8 for the path of the firm's share price. The next day
the firm announced further losses. Borrowing became well nigh impossible
for Lehman. By the 15th, it had filed for chapter 11 bankruptcy protection.
This was one of the largest failures in American corporate history.

Lehman was unusual in that no rescue was arranged. The two government entities were too big to fail: Bear Stearns was too connected. The judgement was that Lehman was neither, and so it could be allowed to go bankrupt.

**Figure 1.8**: The Lehman Brothers share price in 2008

For a few days, this decision seemed harsh but fair. But then the markets realised the implication of Lehman's failure: no financial institution's debt was safe. If you lent to someone, not only might they not pay you back; the government wouldn't either. Given that it was unclear who was running which risks, no financial institution was safe. At that point the argument was decided firmly in favour of the debt market's pessimism, and equity markets plunged.

### 1.4.3  The Various Fates of the Final Three

This lesson had immediate consequences for the other three large broker/dealers. Merrill Lynch, the next most vulnerable firm, quickly sold itself to Bank of America. The final two, Goldman Sachs and Morgan Stanley, turned themselves into banks. This allowed them to access funding from the FED and to be able to fund themselves using retail deposits (something broker/dealers could not easily do).

Meanwhile, the money markets dried up. Financial institutions became increasingly unwilling to lend, either to each other or to anyone else. This began to seriously affect the broader economy. Share prices crashed as the market digested the likely impact of slowing demand and more expensive credit.

### 1.4.4   Bailouts and Busts

The fallout from the failure of Lehman arrived quickly. Washington Mutual was the largest thrift in America. As a prominent loser in the Crunch, WaMu, as it was known, was already looking vulnerable. A downgrade of its credit rating on the 15th September raised concerns further. WaMu found deposits being withdrawn. The resulting pressure on funding caused WaMu's regulator to act. On the 26th September 2008, the Office of Thrift Supervision took control and appointed the FDIC as receiver. WaMu's branches and assets were quickly sold to JPMorgan[24]. An institution with total assets of over $300B had failed in less than two weeks.

It was clear at this point that dramatic action was needed to prevent a string of bank failures. Not saving Lehman might have been a good decision ethically, but it began to look like very bad economics. Many financial institutions were vulnerable, and even the largest banks were having problems raising funding[25]. Government intervention was needed across the whole financial system.

The U.S. Treasury suggested a program to purchase distressed assets from financial institutions, the *Troubled Asset Relief Program* or TARP. This did not meet with favour from legislators, not least because the original plan gave extraordinary discretion to the Treasury with rather little oversight. The House of Representatives voted down the first version of the TARP, presumably taking the view that if the Treasury bought troubled assets for their fair value then it would not have done any good, but if it bought them for more than they were worth, then it amounted to a simple subsidy to the banks who had taken the biggest risks.

The TARP was restructured to include extra oversight and require that the Treasury obtain equity stakes[26] in the firms that they assist. This at least involved less moral hazard than buying assets, and the new TARP was passed on the 3rd October 2008[27].

The program was immediately put to use. Over $125B of new capital was injected into leading American financial institutions. At the same time, other governments were intervening to save their banking systems. For instance, the UK government enacted a £500B plan[28]; Fortis was rescued by the Dutch, Belgian and Luxembourgeois governments; Hypo Real Estate Holding was bailed out by a consortium including the Bundesbank; and all three of Iceland's big banks were nationalised. Figure 1.9 shows some of the larger injections of state capital: in addition a number of banks raised extra capital privately.

| Institution | Amount | Received from |
|---:|:---:|:---:|
| Citigroup | $45B | TARP (twice) |
| AIG | $40B | TARP (originally $80B) |
| Bank of America/Merrill Lynch | $25B | TARP |
| JP Morgan/Bear Stearns/WaMu | $25B | TARP |
| Wells Fargo/Wachovia | $25B | TARP |
| Goldman Sachs | $10B | TARP |
| Morgan Stanley | $10B | TARP |
| Royal Bank of Scotland | £20B | UK |
| Lloyds TSB/HBOS | £17B | UK |
| Hypo Real Estate | €50B | Germany |
| Fortis | €11.2B | Belgium, Netherlands Luxembourg |

**Figure 1.9**: Some recapitalisations during the third phase of the Crunch

The immediate crisis was thereby averted. The recapitalisations and rescues gave the largest financial institutions some breathing space, and a modicum of confidence returned to the market. The Central Banks helped too, flooding the markets with liquidity and cutting interest rates in a coordinated move[29]. Various European nations including Ireland, Denmark and Austria extended the scope of their deposit protection, further bolstering confidence.

Further bank recapitalisations continued through the rest of October 2008: the Swiss banks got more money on the 16th; ING on the 19th; smaller U.S. banks on the 27th. But the pattern was now clear. Governments would bail out their financial institutions by lending them money and injecting extra capital. Shareholders would suffer, but the financial system would not be allowed to collapse.

The position of taxpayers was less clear. A lot of money had been spent by governments in a short period of time. Significant stakes had been purchased for that money in a depressed market. Some of these would prove to be good investments. But there are other assets the taxpayer made which, in the Deputy Governor of the Bank of England's words 'clearly have a level of defaults in them [which we are] not quite sure how will balance out against the residual of the capital'[30]. That is, losses are possible.

## 1.4.5   Qualitative Easing

Interest rates had reached low levels by December 2008. Some Central Banks had very little room for manoeuvre. For instance, rates in the U.S. had been cut to 1% in October, and again, to close to zero, in December[31]. Even the notoriously hawkish ECB had cut rates. The global economy was still in trouble, so a different kind of stimulus was needed.

There were two main answers. Both rely on the Central Bank printing money. The difference is what the money is used for.

- In a *Keynesian Stimulus*, the government spends the money on real assets. Usually this would be inflationary, but with prices collapsing, inflation is a minor concern. The priority instead is to prevent economic collapse by simulating demand, as first recommended by Keynes[32]. Thus the state engages in infrastructure projects, for example, to soak up economic capacity that would otherwise go to waste.

- Whereas in *quantitative easing*, financial assets are purchased. If long term bonds are bought, this keeps down longer term rates. Thus easing ensures that it is not just the cost of short term borrowing that is close to zero, but that of longer term borrowing too. Securities are purchased from banks, freeing up their balance sheets, and allowing them to lend again.

This is the regime we have now entered. Various stimulus programs have been enacted around the world, and more may well be coming. Whether they are enough to steer the global economy out of recession in 2009 remains to be seen.

## Summary

The first phase of the Credit Crunch began in July 2006 when American house prices started to fall. Some areas suffered only minor falls but others, particularly recent and speculative developments in California, Nevada and Florida, fell substantially. Homeowners were either unwilling or unable to pay their mortgages, and so foreclosures rose rapidly. The sale of foreclosed homes depressed prices further.

Many of the affected mortgages had been originated under the originate to distribute model and so their risk had been traded around the financial system. The holders of this risk suffered losses. As the prices of mortgage risk fell and financial institutions lost money, various financial markets were affected. The second phase of the Credit Crunch began as risk capital was

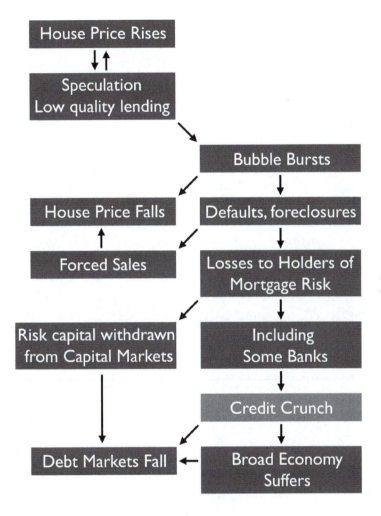

**Figure 1.10**: A brief summary of the Credit Crunch

withdrawn from the financial markets; liquidity fell; and credit became harder to get and more expensive when you could get it. Declining credit quality and increased uncertainty meant that banks became less willing to lend to each other. Central banks tried to help by cutting short rates but this was of only limited help: Libor remained high and so the real cost of funds for banks did not fall substantially.

A number of financial institutions failed or had to be rescued, including Northern Rock, Bear Stearns, Lehman Brothers, AIG and the two American mortgage enterprises, Freddie & Fannie. This shook confidence further, leading to the third phase. Firms reduced their risk by selling assets, further depressing prices. Investors realised that the future performance of many assets – including packages of U.S. mortgages – was uncertain. They became very risk averse. The decline in the amount of risk capital in the markets meant that even if assets were cheap given their likely risks, there were few buyers.

The contraction of credit and declining economic confidence affected the broader economy. Growth slowed in many countries. Equity markets plunged, and governments had to recapitalise the financial system. Partial nationalisations and widespread government guarantees of both of assets and of bank debt followed. What had started as a U.S. housing problem had been transmitted by the financial system into a global economic downturn. This was a direct result of the rules of finance. No one could have planned it.

Nevertheless, the result of many different players optimising their strategies resulted in a recession in the major economies. If we want to understand the result, and how each party was incentivised to behave in a way that was collectively calamitous, we need to look at mortgages, banks, markets, and the framework of rules around each of them. These will be the subjects of the following chapters.

---

## Notes

[1] Some of the material in this book is based on an earlier article in the journal Quantitative Finance. See my paper *A preliminary enquiry into the causes of the Credit Crunch* (Volume 8, 2008).

[2] For more details on the methodology for calculating the S&P/Case-Shiller® Home Price Indices and their history, see www.standardandpoors.com.

[3] This is controversial: many authors would put it a year later, when the problems became clear. However since the housing market is a key cause of the Crunch, the peak in house prices seems a natural starting point. After that things may have looked rosy in the financial system for some months, but trouble was inevitable.

[4] See www.realtor.org/research/research/metroprice for detailed geographical information on U.S. residential house prices, and *The State of the Nation's Housing 2008*, Joint Centre for Housing Studies of Harvard University for a comprehensive account of the housing market.

[5] See *Population Profile of the United States*, U.S. Census Bureau, available at www.census.gov/population/www/pop-profile/files/dynamic/PopDistribution.pdf.

[6] In the United States, mortgage foreclosures are broadly a matter of state law, and the differences between states are considerable. For instance in New Hampshire, foreclosures do not usually involve the courts. Here typically the lender must send a default notice to the borrower, wait thirty days, then a notice of sale can be posted. Three weeks after that, if the default

has not been cured, the property can be sold. In contrast, in New York court proceedings are usually required, and the foreclosure process often lasts over a year. The details are complex: see for instance Rao et al.'s *Foreclosures: Defenses, Workouts and Mortgage Servicing* for more information.

[7] Press release *Foreclosure activity up 14 percent in second quarter*, July 25th 2008 available on `www.realtytrac.com`.

[8] See the Federal Reserve Statistical Release *Charge Off and Delinquency Rates on Loans and Leases at Commercial Banks* available via `www.federalreserve.gov/releases/charge off`.

[9] Privately-owned housing starts were at 965,000 in July 2008 according to the U.S. Department of Housing and Urban Development. See `www.census.gov` for more details.

[10] The NAHB/Wells Fargo Housing Market Index provides information on sentiment within the residential house construction industry. See `www.nahb.org/page.aspx/category/sectionID=134` for details.

[11] The importance of housing to the U.S. economy is well illustrated in *Housing is the business cycle* by Edward Leamer, available at `www.kc.frb.org/publicat/sympos/2007/PDF/Leamer_0415.pdf`.

[12] For more information see the Central Statistics Office Ireland website at `www.cso.ie`.

[13] See the Bank of Spain's *Financial Stability Review* April 2008 available at `www.bde.es/informes/be/estfin/completo/estfin14e.pdf`.

[14] In practice of course regulators would intervene well before that point.

[15] It is also familiar to economists: George Akerlof won a Nobel Prize for his work in this area. His original paper is *The Market for 'Lemons': Quality Uncertainty and the Market Mechanism* in the Quarterly Journal of Economics, Volume 84, Number 3, 1970.

[16] The ABX Indices are produced by Markit: see `www.markit.com` for more details.

[17] U.S. depositary institutions are required by law to maintain certain levels of reserves, either as reserves with the FED or as vault cash. The amount required depends on the size of the institution's deposits, specifically their demand deposits. There is an active market in these FED funds: banks which do not have sufficient FED funds can borrow from institutions with a surplus. Actual FED funds is the rate for these overnight borrowings. The short rate is the FED's target for actual FED funds: typically the actual rate is within a small spread of the target, so for instance on the 8th September 2008 actual FED funds was 1.92% versus a target of 2%.

[18] The conventional view is that this risked inflation: presumably the FED took the view that preserving the health of the U.S. banking system was more important than the risk of higher prices.

[19] Strictly, to make a fair comparison, one should compare rates of identical term. FED funds is an overnight rate so comparing it with three month Libor also involves (a small) term structure of interest rates. Thus more detailed comparisons are based on the expected one month average of the FED funds rate – known as the *overnight index swap* or OIS rate – and overnight Libor.

[20] Broker/dealers are a type of financial services firm unique to the United States. Typically they have limited retail banking and derive most of their profit from their activities in the capital markets. As a result of this they are regulated differently from banks: see chapter 7 for more details.

[21] One of the largest collapses that we do not discuss is that of the Californian Bank Indymac. See `www.fdic.gov/bank/individual/failed/IndyMac.html` for more details.

[22]These are discussed further in chapter 8.

[23]This is perhaps unfair: the U.S. public debt at the time of the conservatorship was just shy of $10 trillion, and Fannie and Freddie's liabilities would add more than $5T more to that. The body responsible for the Federal Budget, the Congressional Budget Office, took the view that 'Fannie Mae and Freddie Mac should be directly incorporated into the federal budget.' (See *Cost of US loans bail-out emerging*, Financial Times, 9th September 2008 available at www.ft. com/cms/s/0/e30472a6-7e79-11dd-b1af-000077b07658.html for more details.) The Bush administration however disagreed, presumably viewing a one day jump in public debt from less than ten to more than fifteen billion dollars as not to their credit. Official statements of the deficit therefore remain unencumbered by Fannie and Freddie's liabilities.

[24]See OTS press release OTS 08-046, *Washington Mutual Acquired by JPMorgan Chase* 25th September 2008 available via ots.gov.

[25]See for instance Alistair Darling's comments on RBS in BBC's *Panorama* of 22nd December, a transcript of which is available at news.bbc.co.uk/1/shared/bsp/hi/pdfs/ 221208_darling.pdf.

[26]Specifically warrants which give the Treasury the right to purchase non-voting stock.

[27]The TARP was enacted as part of the Emergency Economic Stabilization Act of 2008. The full text of the act can be found at banking.senate.gov/public/_files/latestversion AY008C32_xml.pdf.

[28]See *Statement by the Chancellor on financial stability* available at www.hm-treasury.gov. uk/statement_chx_081008.htm.

[29]See ECB Press Release *Monetary policy decisions*, 8th October 2008, available at www.ecb. int/press/pr/date/2008/html/pr081008.en.html.

[30]See John Gieve's comments in BBC's *Panorama* of 22nd December, a transcript of which is available at news.bbc.co.uk/1/shared/bsp/hi/pdfs/221208_gieve.pdf.

[31]See the FED's Press Release of 16th December, setting a target for FED funds of between zero and a quarter of one percent. This is available at www.federalreserve.gov/ newsevents/press/monetary/20081216b.htm.

[32]This is, of course, a caricature of a complex argument. See Paul Krugman's *The Return of Depression Economics* for a more extensive discussion.

# Chapter 2

---

# Understanding the Slime: U.S. Residential Mortgages

> *Men are but children of a larger growth;*
> *Our appetites as apt to change as theirs,*
> *And full as craving too, and full as vain.*
> John Dryden

---

## Introduction

The Credit Crunch was triggered by mortgages: the risks of mortgage lending were the root cause of many of the losses experienced during 2007 and 2008. Therefore we need to look in detail at U.S. mortgages: how they are structured; who could borrow; and who could lend. This is important because it is not just falling house prices that caused the Crunch. Rather it is a combination of falling house prices and risky lending. Other countries have suffered house price falls recently (as we saw in the previous chapter with Spain and Ireland). But their financial systems – while distressed – have not gone into meltdown as a result. This is because European mortgage structures were not as risky and European mortgages were not given to such a high proportion of borrowers who would not or could not repay them[33].

A good example of the toxic nature of American home loans can be found in the trajectory of Countrywide Financial. This was a very large U.S. mortgage lender based on the ODM. We show how it was able to grow very fast during the Boom years, and how this growth inevitably lead to disaster when the house price bubble burst.

## 2.1 Mortgage Structures and Borrowers

The structure of a mortgage determines the monthly payment needed for a given house, and hence who can afford to buy it. Since the 1980s, various mortgage structures have been created which lowered payments, making home ownership more and more affordable. However in many cases this was at the expense of increased risk to the borrower, the lender, or both. In this section we look at some of these innovations and how they changed the nature of the mortgage market.

Another change since the 1980s concerned who could get a mortgage. Slowly over the 1990s, mortgages became accessible to a wider and wider range of people in the U.S., some of them of poor credit quality. Even something as extreme as the *Ninja* mortgage – one given to people with *No Income, No Job* and no *Assets* – became commonplace. This helped to stimulate the property price boom of 2000–2006 as more and more people could buy a home. But the combination of riskier loan structures and lower credit quality borrowers left the holders of mortgage risk with much bigger losses when house prices turned down.

### 2.1.1   Basic Mortgages

A standard mortgage in the U.S. has the following structure:

- It is a loan of a fixed sum of money known as the *Notional* amount;

- The loan is for a *Fixed Term*, often thirty years;

- The mortgage is *Prepayable*, that is the borrower can at any time pay back the current mortgage balance and terminate the loan;

- Payments are due *Monthly*; and

- Payments are fixed for the whole term of the mortgage. If these payments are made every month for the term of the mortgage, then the mortgage will end with nothing left to pay. At this point the borrower owns the property outright.

This is called a *Level Pay* structure, and most residential mortgages were of this form by the 1980s[34]. Here, the payments at the start of the loan are mostly interest, with only a small fraction going to reduce the principal balance. But as the principal balance is slowly eroded, less interest is needed and so the principal balance declines faster later in the mortgage.

Consider a typical $200,000 thirty year level-pay loan. The monthly payment is roughly $950, and this is split into a payment of interest, a repayment of the principal balance, and an administrative fee to the servicer. Figure 2.1 shows how the proportion of the monthly payment going to each source varies over time.

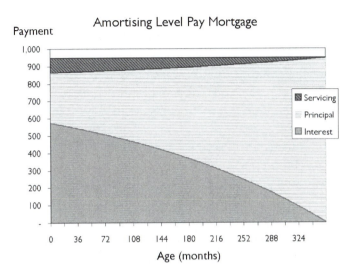

Figure 2.1: The components of a thirty year level pay mortgage over time

Prepayment is an important feature of the mortgage in that it provides flexibility: if a borrower wants to move, they can sell their property, prepay the mortgage, and get a new loan for their new home. Similarly if property prices are rising, a borrower can *refinance* by prepaying their mortgage and taking out a new one for more money. This allows them, in the words of Dave Simonsen, to use 'their house as an A.T.M. machine' or, in the jargon, *cash out refi*[35]. This feature increases the feedback between house prices and the general economy discussed in the previous chapter.

### 2.1.2 *Mortgage Collateral and LTV*

What is the risk of a mortgage to the lender? One dimension concerns the principal balance of the loan versus the value of the house. If the house is worth $250,000 and the borrower takes out a conventional level pay mortgage with a notional of $200,000, even if they never make a mortgage payment, the lender still has a cushion of $50,000. Suppose that the foreclosure process takes a year. In that case, the lender might well judge that it is unlikely that the house will have dropped by that much; thus they will get enough in a foreclosure sale to repay the loan, and so the mortgage is reasonably safe for the lender.

The *Loan To Value* Ratio (LTV) of a mortgage is simply the ratio of the loan notional to the value of the property. Thus in our example it is $200/250$ or 80%. Traditionally, lenders were unwilling to lend above an 80% LTV as it did not give them enough safety in the event of default by the borrower. Certainly if they did lend at higher LTVs, then the interest rate on the mortgage would be considerably higher to provide compensation for this extra risk.

LTV is a less significant issue in much of continental Europe than in the United States. This is due to the different nature of the claim a lender has in Europe versus the U.S. In Europe, most lending gives the bank recourse against the borrower's other assets if they default on their mortgage and the property is not sufficient to repay the loan. For instance the bank can seize the borrower's car. This means that defaulting on a mortgage is the last resort for European borrowers. Moreover it is in their interests to ensure that the value of the property is maximised in foreclosure[36].

In many U.S. states however, *non-recourse* lending is either common or required. Here the bank can only take the property and nothing else. This sets up an entirely different incentive. First, if house prices fall – so that the principal balance on the mortgage is larger than the value of the property – then there is no incentive to carry on paying the mortgage. You simply give the property to the lender and walk away[37]. Second, there is no incentive to maximise the value of the property in foreclosure: rather the reverse, there is incentive to take and sell anything that the borrower can remove. This leads to foreclosed homes lacking doors, windows, lighting, kitchen units or plumbing as discussed in chapter 1[38].

These risks only really become apparent in a falling house price environment. When times are good, the longer a mortgage is held, the lower risk it is. This is because rising house prices lower the effective LTV: after a few years even a mortgage that was originated at 100% LTV will be backed by a house worth much more than the mortgage amount. Therefore net losses on mortgage defaults are small. It is only when prices start going down that borrowers have an incentive to default. Default rates rise just when effective LTVs are going up too. This can be a toxic combination particularly if the portfolio of mortgages concerned has a lot of recent mortgages (so that previous periods of house price inflation have not had a chance to lower LTVs).

### 2.1.3   Mortgage Affordability and PTI

Borrowers only pay their mortgages if they can afford to do so. One measure of affordability compares the monthly payments on debt with the borrower's monthly after-tax income. This payment-to-income or *PTI* ratio is typically limited by lenders at 40% or so. Certainly if debt repayments take up more than 50% of a typical borrower's income then it is likely that the mortgage will be unsustainable and default is more likely.

At any given moment the monthly payment on a new thirty year level pay mortgage depends on long term interest rates. The lender has to borrow for a long period in order to fund the mortgage, so they can only afford to charge a mortgage rate which reflects their cost of borrowing, plus the cost of servicing the mortgage, plus some compensation for default, plus some profit. The cost of funds component means that if long term interest rates go up, mortgage rates go up, and borrowers who could have got a $200,000 mortgage for $950 a month suddenly find that if they have not signed up quickly enough, they now have to pay $1000 a month or more.

Similarly falling rates reduce the payments on new level pay mortgages. In a low interest rate environment, then, level pay mortgages are more affordable than when rates are higher. Falling rates gives rise to another source of prepayments: mortgage holders whose payments were based on previous, higher rates see that they can refinance into a cheaper mortgage and so they prepay their existing loan and enter into a new one, thus reducing their monthly payment. Thus mortgage portfolios prepay faster when interest rates are falling and more slowly when rates are rising.

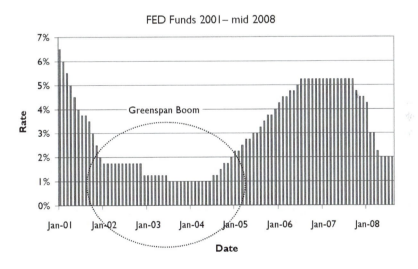

**Figure 2.2:** FED Funds and the Greenspan Boom

One of the major drivers of the U.S. house price boom which ended in 2006 was low interest rates. Alan Greenspan, as Chairman of the FED, cut interest rates very aggressively after the 9/11 tragedy, as figure 2.2 illustrates. This gave rise to the *Greenspan Boom* as low borrowing costs caused stock markets, bond markets, and other asset markets to move up and up. Investors borrowed cheaply to speculate. Property markets in particular went up as more and more prospective borrowers found a mortgage they thought that they could afford.

### 2.1.4   ARMs and Resets

Typically the cost of borrowing for shorter periods is less than that for borrowing for longer periods: most of the time the *yield curve* (a plot of the interest rate for borrowing for a given period against that period) points up. This means that a mortgage based on short term interest rates is likely to be, on average, cheaper than level pay structure.

*Adjustable Rate Mortgages* exploit this fact. In an ARM, the interest rate resets regularly based on the current level of some reference rate. Typically the resets are annual. The most common reference rates are:

- 12 month Libor;
- One year Treasury rates; or
- An index known as COFI which reflects the cost of raising money for medium sized banks[39].

Suppose a borrower takes out a thirty year ARM based on 12 month Libor. Every year at the reset date the lender observes Libor. Mortgage payments for the next year are based on that rate plus some fixed *spread*: thus for instance the payment might be one year Libor plus 1.5%. If Libor goes up from one year to the next, mortgage payments rise too; while if Libor falls, so do the payments for the next year. Thus the borrower has the risk that interest rates will rise over the term of the mortgage[40].

Figure 2.3 illustrates the key rates for ARMs and conventional level pay mortgages: for an ARM, only the twelve month rate matters, whereas payments on level pay mortgages are also affected by much longer term rates[41]. For the yield curve shown – which reflects the average shape for the U.S. dollar Treasury yield curve between 2000 and 2005 – ARMs offered a cheaper headline rate than level pay mortgages. This reduction in cost reflected the interest rate risk that the borrower was taking on: the risk that the reference rates would rise.

### 2.1.5   Interest Rates and Prepayment

Notice that a borrower in a level pay mortgage has all of the advantages of falling interest rates – in that they can refinance if it is favourable to do so – but none of the disadvantages. They pay for that privilege by having a higher mortgage rate[42]. An ARM restores the symmetry between borrower

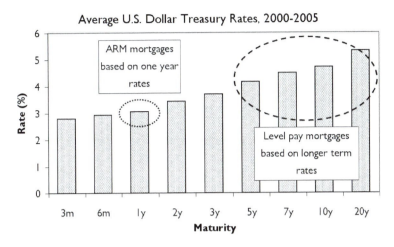

**Figure 2.3**: The average U.S. Treasury yield curve, 2000–2005

and lender: the borrower benefits from lower rates; but they suffer from higher ones. This too helps make ARMs cheaper than level pay mortgages. Thus while many ARMs are prepayable, it typically does not make sense for the borrower to prepay simply because interest rates have fallen: they will receive the advantage of those lower rates at the next reset.

ARMs make life easier for lenders too. The problem with a level pay mortgage is that the lender does not know how long to fund the mortgage for. If we have a thirty year mortgage, then the borrower could be repaying principal for the entire thirty years. This suggests that it would be prudent for the mortgage lender to use thirty year funding. But if interest rates fall, then many mortgages will prepay, leaving the lender with expensive liabilities – its borrowing – but no assets. It cannot replace those assets with new mortgages at the same rate due to the fall in interest rates. The only ways to deal with this problem are:

- Either estimate how many prepayments there will be for a given move in interest rates, calculate how much that will cost the lender, and buy interest rate derivatives to hedge that risk; or

- Make it somebody else's problem by selling the mortgage.

ARMs do not suffer from the same sensitivity of prepayment to interest rates and hence they are potentially more attractive to holders of mortgage risk who do not wish to take *prepayment risk*.

The incentive to refinance a level pay mortgage depends on the difference between the rate on the mortgage and the current mortgage rate. Thus if I

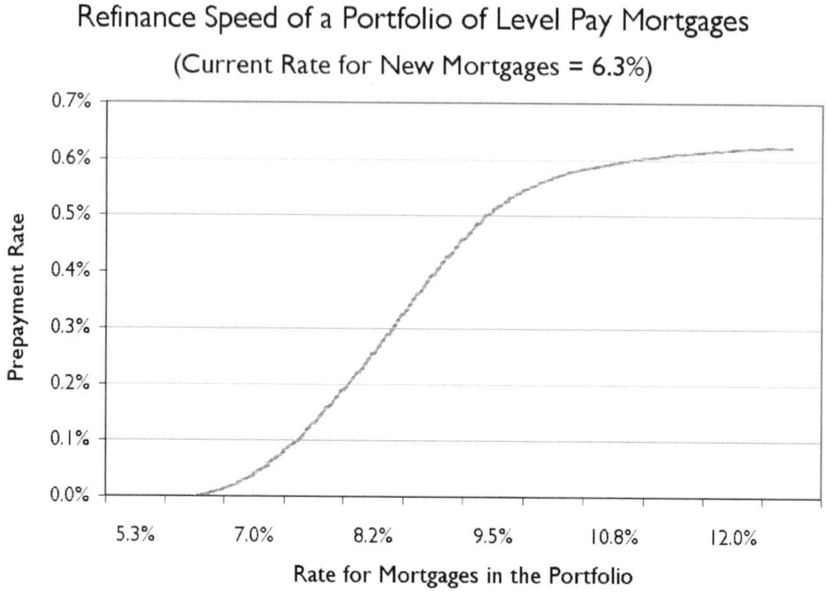

Refinance Speed of a Portfolio of Level Pay Mortgages

(Current Rate for New Mortgages = 6.3%)

**Figure 2.4:** The interest rate sensitivity of prepayment rates

took out a 10% mortgage when rates were high, and current rates are only 6%, I have a significant incentive to prepay the mortgage and refinance into a cheaper loan. On the other hand, if my mortgage rate is 6.5%, then my monthly payments would not go down nearly as much on a refinance, and hence my incentive is smaller. Figure 2.4 shows typical prepayment rates as a function of the original mortgage rate with current rates at 6.3%.

### 2.1.6   Hybrid ARMs

Suppose borrowing under an ARM is still too expensive to allow you to afford the property you want. How might you structure a mortgage to make the payments even lower?

One approach is an *interest only* mortgage. Here you do not reduce the principal balance at all: you just pay interest on it. This reduces payments to some extent. If the borrower takes the view that house prices will go up, then using an interest only mortgage will allow them to buy a more expensive house. In time the house can be sold for more than its purchase price, so the mortgage principal will be paid off at that point. If prices go down, however, the borrower will not be able to repay the mortgage and the lender is likely to suffer a loss, especially if the loan is non-recourse.

*Hybrid ARMs* also give low mortgage payments, at least at first. Here the lender offers a fixed rate for a period at the start of the mortgage, typically between one and five years. This initial rate is known as a *teaser* as it is typically set fairly low. During the teaser rate period, the mortgage payments are not sufficient to repay much if any principal on the loan. Indeed they may even be *less* than those required on an interest only loan: in that case the principal balance increases during the teaser rate period. These *negative amortisation* products are particularly risky for the lender as the amount that they need to recover from a property sale in foreclosure increases during the early years of the mortgage.

### 2.1.7    Option ARMs

Borrowers like to have flexibility. If they are doing well they might like to increase their mortgage rate and so pay off the mortgage faster (or at least avoid negative amortisation); whereas if their income falls, they may wish to reduce their mortgage rate temporarily. An *option ARM* gives the borrower the choice between several different mortgage rates during the initial period. Thus every year the borrower might have the option to pick one of three payments. The lowest, minimum payment makes the mortgage most affordable but it produces negative amortisation. The next one is an interest only payment, while the highest one is the same as the payment on a thirty year level pay mortgage.

Negative amortisation cannot go on forever if lenders have any degree of prudence. Suppose we start with a 90% LTV mortgage on a $200,000 property, i.e. the initial mortgage principal is $180,000. The lender may decide that they do not wish the mortgage amount to rise above an effective 100% LTV. Thus the borrower only has the choice to pick the lowest rate if the principal balance is less than $200,000: once negative amortisation has caused the principal to grow to that level, they can only select from interest only or level pay rates. The option ARM therefore has a *negative amortisation cap*[43].

Option ARM products were also sometimes found with an initial low teaser. This was sometimes as little as 1%, so option ARMs often allowed borrowers to buy homes which would have been completely unaffordable under a level pay structure. Figure 2.5 illustrates this. It shows how expensive a house a borrower able to spend $1,000 a month on a mortgage could afford using various mortgage structures based on the rates at the start of 2005 and a 95% LTV.

Products with low teaser rates encourage speculation in a rising house price environment. If house prices go up 10% and I buy the most expensive

| Mortgage structure | 30 year level pay | ARM | 1% teaser rate option ARM |
|---|---|---|---|
| Max price affordable | $220,000 | $305,000 | $1,250,000 |

**Figure 2.5**: Approximate maximum house price for various mortgage structures based on typical rates in early 2005 and a $1,000 per month mortgage payment

house I can afford using a 30 year level pay structure, I make $22,000. But if I do the same thing with a 1% teaser option ARM, I make $125,000.

### 2.1.8  Borrower Quality and FICO

Who should lenders allow to borrow? One might have thought that a proven history of failing to repay debts would make it harder to get a mortgage, and until 2000 or so this was the case.

There are various ways of measuring the credit quality of an individual. *Credit scores* typically look at various factors including:

- *Speed of payment of previous debts*: any delay in paying debts such as credit card bills or utility bills lowers the credit score as, of course, does any failure to pay;

- *Amount of debt taken up*: high use of available credit card or overdraft facilities lowers the credit score;

- *Length of history*: the more history is known about the borrower, the better;

- *Amount of debt recently applied for*: applications for new credit are typically used as a proxy for the borrower's need for credit, so more is worse from a credit scoring perspective.

One of the best known credit scores is *FICO*[44]. FICO scores are available on many individuals within the U.S. They range between 300 and 850.

Different lenders have different thresholds for FICO scores. Typically the best quality borrowers are regarded as having a FICO above 690 or so. Between 690 and roughly 640, borrowers are seen as lesser quality but still fairly good: a score below 640 typically indicates a higher risk borrower.

One of the major features of the Greenspan Boom was increasing availability of credit to low scoring borrowers. This is partly because rising house prices made this form of lending look low risk: when borrowers could buy a million dollar house, pay less than a thousand dollars a month on their mortgage, and flip it a year later with a hundred thousand dollar profit, they tended to keep paying the mortgage.

### 2.1.9    Prime, Alt-A and Subprime

What would be a good mortgage from the lender's perspective? A reasonable list would be:

- It would have an LTV below 80%;

- The borrower would have a high FICO score, certainly above 650;

- The PTI ratio was below 40%; and

- The borrower would be able to prove their level of income so we were confident that the PTI ratio was correct.

The term *prime* is used to refer to good quality loans. It is not a precise term – some lenders might consider a 625 FICO borrower prime if their LTV and PTI were good enough – but it implies some measure of safety in LTV, FICO and PTI.

There are a number of reasons a borrower could fail to be prime, some of them benign. For instance a self-employed worker with a high but variable income might not meet the documentation requirements to prove their income to the lender's satisfaction. Alternatively a high credit score borrower might want a loan that is easily affordable on a PTI basis but with a 95% LTV. These borrowers are known *Alt-A*: they are sometimes high quality borrowers, but they do not meet all of the criteria to be prime.

Below Alt-A comes *subprime*. Here we find borrowers with a poor credit history, perhaps due to recent delinquencies. The term is again imprecise, with *stated income* loans – where the borrower simply asserts what they earn without being able to prove it – are sometimes called Alt-A and sometimes subprime. Once a borrower can prove neither income nor job status, as in a Ninja loan, we are definitely in subprime territory. Typical features of a subprime borrower include:

- At least one failure to pay over the last year (technically, one or more 30-day delinquencies); or

- An action of foreclosure or repossession against the borrower in the last two years; or

- Bankruptcy in the last five years; or

- A FICO score below 640; or

- A PTI above 50%.

The effect of borrower quality on mortgage rates is easiest to see by looking at the spread of ARMs. The typical spreads paid by various types of one year Libor based ARM are shown in figure 2.6. Lenders are compensated for the extra credit risk of subprime borrowers by between one and six percent extra interest per year.

| Prime Borrower | Alt-A | Subprime |
|----------------|-------|----------|
| 2.25-2.75% | 2.5-3.5% | 3.25-7.5% |

**Figure 2.6**: Typical ARM Libor spreads for different borrower types during the Boom

Subprime loans, then, were often expensive for the borrower. While house prices were rising, they were relatively low risk for the lender.

### 2.1.10   Houses as ATMs: Second Mortgages and HELOCs

Suppose that you own a house bought using a mortgage. If house prices have gone up, then there are two obvious ways to extract the gain and turn it into cash:

- Repay your current mortgage and refinance into a loan with a larger notional; or

- Take out a second mortgage[45].

The first method – cash out refi – is only attractive if the rate on the new loan is lower than that of the old one. The second approach gives rise to a loan known as a *home equity loan* or HEL.

A variant on the HEL is the HELOC or Home Equity Line Of Credit. Here the lender sets a maximum amount that can be borrowed but does not advance all of the funds at once. Instead the HELOC is structured like a credit card: funds can be borrowed up to the limit, repaid, and reborrowed again as the mortgage holder wishes providing that they continue to pay the agreed interest rate[46] on the amount outstanding.

Many borrowers like the flexibility that HELOCs provided. Some kept the HELOC for emergencies, others used it sparingly to spread the costs of big purchases. And of course some spent it all at once on consumer goods. HELOCs became broadly popular, and some first mortgages were even structured with a HELOC feature.

*Summary*

The character of mortgages changed dramatically during the ten years before the Crunch. Mortgage lending used to be a low risk activity: LTVs below 80%, PTIs below 40% and FICOs above 650 all protected lenders. Moreover, with a standard level pay loan, the mortgage became safer after a few years of payments as the principal balance declined and, thanks to house price inflation, the collateral value rose. Moreover borrowers had certainty of their outgoings over the whole term of the mortgage and so could plan ahead with confidence.

All of this changed during the Boom years. Higher LTVs and lower FICOs became common. PTI was still important, but it was based on the low teaser rate rather than the eventual mortgage payment after that rate had expired. This allowed borrowers to buy much more expensive properties. If the property went up in value during the teaser period, they made a lot of money: if it went down, they defaulted on the mortgage. Non-recourse structures were truly 'heads I win, tails you lose' games for the borrower.

Moreover negative amortisation meant that some mortgages got riskier as time passed, while easy HELOC availability encouraged those home owners with low LTVs to borrow more. The growth of the Alt-A and subprime markets changed the character of borrowers, particularly during the last years of the bubble. Mortgage lending was no longer a low risk activity.

## 2.2 How Mortgages Were Made

Many loans were made during the Greenspan Boom that would not have been seriously contemplated a few years earlier. This dramatic expansion of lending was caused by incentives in the mortgage market to keep lending. It is time to look at the players in this market and how they were paid.

### 2.2.1 *Mortgage Brokers*

The naïve picture of mortgage origination is that the borrower walks into their local bank branch, fills in a few forms, and waits for a mortgage decision. This is still a common way for mortgages to be made in some European countries, but the picture is more complicated in the U.S.

First *mortgage brokers* are common in America. A broker assesses a potential borrower's circumstances, suggests a lender and a loan type that might be suitable for them, and assists in the loan application process. Many realtors either have captive brokers or brokers they usually suggest. Some individuals develop a relationship with their broker and come back to them every time they need a new mortgage. In addition brokers market their services actively. All of this combines to give brokers a significant rôle in mortgage origination: in 2006, for instance, they introduced lenders to borrowers in over half of American residential mortgage loans[47].

The ubiquity of mortgage brokers means that having bank branches is not enough to ensure significant loan origination volume: in order to be competitive in the mortgage market, a bank needs to offer loans that the brokers will recommend. Brokers are paid by the lender for each mortgage that is made. Therefore they have an incentive to ensure that the mortgage *is* made: and this means that they are encouraged to send in mortgage applications to the lender that present the borrower in the best possible light. Because they know the lending process well, brokers can 'optimise' applications to make it more likely that the loan will be made, and they can select a lender who will take the best possible view of the borrower's circumstances.

The broker's fee is sometimes a function of the loan type, and they are typically higher for the more complex loan types like option ARMs[48]. This in turn means that brokers are sometimes incentivised to recommend more complex loan types. Finally, brokers are lightly regulated, so their conduct in this area is often less subject to scrutiny than bank lenders.

### 2.2.2    The Lender and Servicing

The lender is the party whose name appears on the mortgage. They provide the money which allows the property to be bought: this is sometimes called *funding* the mortgage.

The *servicer* of the mortgage is the party who collects payments, records mortgage balances and other information about the borrower, and follows up delinquencies. They also typically have some responsibility for negotiating loan modifications for borrowers who are having trouble paying their mortgage. If a delinquency cannot be cured, they are responsible for the foreclosure process. Finally they may also remit the borrower's property taxes and property insurance.

Servicers are paid a servicing fee for their labours. This is typically a percentage of the interest paid on a loan, and the percentage again varies by loan type. 0.25% for level pay mortgages and 0.375% for ARMs is common: it may well be higher for subprime loans. Again this means that servicers prefer more complex loan types and so if they are also direct lenders, they will tend to market these loan types more actively.

Servicing is a competitive business[49]. If it costs too much, then the lender cannot offer competitive loan rates. Moreover there is little incentive to do it well since the borrower cannot change servicer. Therefore there is a temptation for servicers to be under-resourced, or at least only just well enough resourced to handle the work they have. This in turn means if there is a spike in delinquencies, servicers may not be adequately resourced to chase all of them up. If servicers are paid for the extra costs of high delinquency rates, then they will be encouraged to be diligent; to follow up on delinquencies; to structure modifications which reduce loss rates where possible; and to foreclose quickly and sell the property where modifications are not possible. But often they are paid a fixed fee regardless of delinquency rates.

None of this would matter if the servicer was also the long term holder of the risk. But this is rarely the case. Instead the initial lender often sells the loan on, often keeping the right to service it. Servicing provides fee income, a key component of the ODM. Here the lender has no incentive to work harder than necessary to earn their fee.

### 2.2.3  Mortgage Funding and Mortgage Banks

Where does the money to fund a mortgage come from? If the lender is a bank, then they have deposits[50] and they can use that source of money to fund mortgages. This is the old-style model of retail banking: take deposits and use the money to make loans. However deposits have not been adequate to support the demand for mortgages for some time.

*Mortgage banks* grew up to meet this demand. These are institutions which specialise in mortgage lending. Despite the name, a mortgage bank does *not* have a banking license so it cannot take deposits. Instead a mortgage bank raises short term funds in the capital markets and uses those funds to make mortgage loans. Once a mortgage has been made, it is packaged up with other mortgages. Portfolios of loans are then sold on, freeing up cash for further lending. The mortgage bank is thus a pure new style institution: without deposit funding, it has to be.

### 2.2.4  Correspondent Lending

Suppose the mortgage market consists of fifteen and thirty year fixed rate loans, various forms of ARM, hybrid ARM and option ARM, and a variety of other loan types. A small depositary bank or mortgage bank may find it difficult to offer its own version of each of these loans. Also the buyers of portfolios of loans will want a standard product as it is easier to assess. The small lender can deal with this problem by entering into an agreement with a larger institution. They 'rebadge' the larger lender's products and sell them as if they were their own. The larger institution agrees to buy these loans, thus solving the lender's funding problem and bulking up their own loan book.

This practice is known as *correspondent* lending. The correspondent is the initial lender: the larger institution is sometimes known as an *aggregator* as they buy loans from a number of correspondents. Aggregators are basically contracting out mortgage origination – including the assessment of borrowers – to their correspondents.

Correspondent lending allowed institutions with very little capital to become mortgage lenders. All that was needed was a relationship with brokers, a correspondent lending agreement with an aggregator, and a small amount of cash to fund the loans in the period between the initial advance and the sale to the aggregator. Servicing capability was not even necessary: you can pay a specialist servicer to do that for you.

Technology accelerated this trend. Lenders could set themselves up to offer loans via a website or over the phone, removing the need for expensive branch offices and lowering their costs. This blurred the distinction between brokers and lenders: some brokers negotiated a pre-approval agreement with aggregators allowing both parties to streamline the origination process.

## 2.2.5   *Who Decides? The Changing Rôle of the Loan Officer*

In the old days, when you walked into the branch of a depositary bank to get your mortgage, you had to face the *loan officer*. This was the person responsible for assessing your mortgage application and deciding whether to advance the funds. They would have had certain criteria set by the bank – of LTV, PTI, FICO and so on – but also a measure of discretion. Good loan officers protected the bank's interests, and judging loan applications was a skilful job requiring experience and training.

As the ODM developed it became less necessary to have good loan officers simply because the consequences of not having them were not suffered by the lender. Why go to the trouble of figuring out whether a second hand car is good or not when you can sell *any* car on for a profit? Similarly why go to the trouble of carefully assessing a mortgage application if you are just going to sell the loan on to an aggregator in a few months and the aggregator appears not to care?

The rôle of loan officer therefore migrated in many cases into a sales person. Their discretion was reduced and the pure numbers such as FICO became more important in loan decisions: indeed some lenders automated the whole process, and let their computers make loan decisions based on forms filled in online. By 2005 if you – or your broker – could figure out how to get a 640 or better FICO, 90% LTV and 40% PTI based on the initial rate, you could almost certainly find someone to lend to you as a prime borrower. It probably helped if you could sign your name too, but that may not have been strictly necessary.

## 2.3   Mortgage Lending During the Greenspan Boom

Low interest rates and teaser rate mortgage structures stimulated massive demand for mortgages during the Greenspan Boom years. By this stage mortgage origination had been transformed from an old-style banking business into a fractured landscape of brokers, servicers, lenders, aggregators and buyers of loan portfolios. Many parties in this process simply wanted to get as many mortgages as possible agreed. This might have been expected to lead to bad loans, and it did.

### 2.3.1   *Where did it all go wrong? Incentive Structures in the Mortgage Industry*

Suppose we have a house buyer being wooed by a realtor. The realtor wants to sell one of the properties they list. To further the sale, they introduce a friendly broker. The broker wants to earn their fee so they find an affordable mortgage – which typically means a teaser rate structure – and they fill in the mortgage application for the buyer. The mortgage is made by a mortgage bank who immediately sells the loan and the servicing rights on to an aggregator. The aggregator keeps the servicing and combines the mortgage with many other loans. This portfolio is sold on to a broker/dealer, who sells it on to an end investor[51].

No one in this chain is incentivised to make sure that the borrower is able and willing to make the payments on the loan in the long term, that the property is worth what the loan application claims it is, or that the borrower understands the risk that they are taking. Interests are not aligned. This is the major reason that so many bad loans were made. The people close to the borrower – the broker and perhaps the initial lender – had no interest in the long term performance of the borrower. The person who did – the eventual buyer of the portfolio of loans – was so distant from the mortgage origination that they had little to no control over it or understanding of the precise risks being taken.

Note that the unique American approach to mortgage lending explains in large part why the loss rates on U.S. mortgage portfolios were so much worse than those on Irish or Spanish loans. In all three cases, there was a property price bubble which burst with property prices turning down fast. However loan structures in the U.S. were much riskier, thanks to lack of recourse, high LTVs, and teaser rate mortgages. This, combined with lower quality, badly-vetted borrowers meant that the American residential mortgage problem was more severe than the European one.

What is startling is the lack of buyer beware in all this. The parties further down the chain seem to have taken little interest in what it was exactly that they were buying. This meant that when the game of pass the mortgage stopped, many of them were surprised by what they found themselves holding.

### 2.3.2    Bad Behaviour in the Mortgage Market: Fraud and Related Scams

One set of problems came from laxity in the initial loan process. Some brokers lied when they filled in mortgage applications: others strived hard to present the facts in the best possible light and to gloss over any issues. Brokers or other salespeople used friendly property appraisers to get the 'right' valuation on the property: 'right' in this case meaning the one that would allow them to get the lender to agree to the mortgage[52].

Home buyers were not innocent either: some individuals used easy borrowing in a criminal or at least borderline illegal way. The main types of borrower fraud are[53]:

- *Income misrepresentation.* Low or no documentation loans offer the temptation for borrowers to overstate their income and/or understate their outgoings in order to qualify for the PTI threshold. This led to stated income loans being referred to in some quarters as *liar* loans. A more extreme example of the same basic phenomenon is identity theft where the borrower assumes another identity in order to get the loan agreed.

- *Seller/buyer or appraiser/buyer collusion.* In both of these cases the value of the property is over-stated, allowing the borrower to get extra funds. These are split with the seller, the appraiser or both. If the loan is non-recourse, the buyer can then default, leaving the lender with a property worth much less than the mortgage balance.

- *Status Fraud.* Here buy-to-let loans were misrepresented as loans for owner-occupied houses or construction loans as loans for completed property. Mortgage rates for investment purposes are higher, so some buy-to-let borrowers either lied about existing mortgages or used friendly relatives to make applications. Similarly loans to fund construction are more expensive than those for completed houses, so fraud occurred here too.

- *Multiple borrowing.* Borrowers apply for several loans – perhaps multiple second mortgages – on the same house simultaneously. If the documentation of each loan is subtly different the borrower might be able to bypass systems checks and get several lots of funds before running away.

The extent of mortgage fraud in the Boom years is extraordinary. One estimate is that it increased more than ten fold from 1997 to 2005[54].

Lax mortgage origination practices made much of this possible, or at least easier to execute. It is difficult to look a loan officer with twenty years experience in the eye and lie about your income: clicking the wrong checkbox on a website or saying 'yes' when a friendly broker asks if you earn over fifty thousand dollars a year is easier.

### 2.3.3 *Prepayment penalties and sharing the rewards*

Suppose you make a risky mortgage, charging a high rate of interest. Given the costs of assessing borrowers and documenting loans, you will want to ensure that the borrower cannot refinance too quickly and deprive you of the interest. This is especially true since it tends to be borrowers with improving credit quality who refinance: they can get a new loan at better terms. The solution to this problem is *prepayment penalties*. The lender may impose a penalty for repayment of the loan during the first few years[55], ensuring that they earn a minimum spread providing that the borrower does not default.

### 2.3.4 *Making Slime: The Rise of Subprime*

We have seen that lots of parties were encouraged to ensure that mortgages – any mortgages – were made. More mortgages meant more fees. Brokers, appraisers and ODM lenders all fall into this category. This helps to explain the rise of subprime: there were too many lenders chasing too few prime borrowers, so the lenders had to become less choosy who they lent to. Figure 2.7 illustrates the surge in subprime lending[56]. Moreover stated income mortgages – some of the riskiest of all subprime loans – increased significantly, growing from about 25% to over 40% of total subprime lending[57].

Clearly this rise in lending to a hitherto non-credit-worthy class of borrower increased demand. Subprime mortgages brought more home owners to the market, and so contributed to the house price boom. This was not just

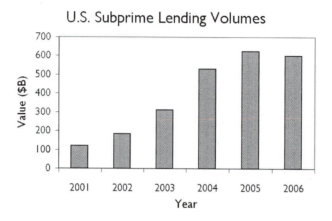

**Figure 2.7**: Subprime mortgage volumes 2001–2006

people who had always been subprime: prime borrowers who fell on hard times and defaulted on their loans had a second chance via subprime mortgages. As the problems with subprime lending became apparent in late 2006 and 2007, the subprime market largely disappeared, and with it a whole class of home buyers. In particular there was no escape route for these 'ex-prime' borrowers. With no subprime loans available, people in this position were now forced to default.

### 2.3.5 Teaser Rate Subprime Loans

Teaser rate loans are classified by the teaser rate period: 2/28 (two year teaser rate followed by twenty eight year floating rate) and 3/27 were common. Often there are prepayment penalties during the teaser rate period[58]. This means that refinancing stress happens exactly at expiry of the teaser rate: at this point, the mortgage rate rises dramatically making the mortgage less affordable; an earlier refinance is unlikely thanks to the penalties. Such a loan behaves rather similarly to a 2 or 3 year interest only mortgage. Most subprime loans were hybrid ARMs with one to three year teaser rates, after which they reset based on one year Libor.

The initial rate on subprime teaser loans – unlike prime teasers – was not particularly low[59]. This compensated the mortgage risk holder for taking subprime exposure. It also gave the mortgage risk holder an implicit two or three year position on house prices: if prices rose, the borrower could refinance the loan, and the mortgage risk holder earned the teaser rate. If prices fell, on the other hand, borrowers would need to make a substantial cash payment in order to refinance. We know that it is likely that they do not have a lot of cash as they are in a teaser product in the first place. So their only option when their teaser rate expires is to default on the loan.

### 2.3.6   The Treadmill of Sales

The strategy of ODM institutions is to originate assets and sell them. Clearly this strategy works best if they can sell the asset for more than its cost. ODM institutions tend to concentrate, naturally enough, on those assets they can sell at a profit. This gives rise to an addiction to the ODM.

To see why, consider an old-style bank following the 3–6–3 rule. At its simplest[60] each year the bank will record profits of 3% of the balance of its loan book. But now suppose the bank moves towards the ODM: it approaches potential buyers for some of its loans. If the average life of the loans is five years, the buyers see a total available profit to them of roughly 15%[61] and so might be willing to pay 10% for the loans. The bank agrees to the deal. It records a *gain on sale* of more than three times its previous annual income.

The bank's profits go up. Equity analysts and investors see this and applaud the gain. But crucially they expect the bank to do it again. The institution has shown that it can be more profitable: the bar has been raised. The only way that the bank can meet this new, higher standard is to originate more assets and sell them. It has been drawn onto the treadmill of origination and sale, and it cannot get off without disappointing shareholders. Fairly quickly originating good assets becomes a distant memory: originating any assets at all is enough providing they can be sold for a profit. Hence the quality of the loan book declines as good assets are sold. The bank starts to become active in subprime mortgage lending.

### 2.3.7   ARM Resets

The rise of subprime gave many borrowers the ability to take a bet on property prices going up with little or no personal risk. By taking out a teaser rate loan which would become unaffordable, they basically bet that prices would rise before the teaser rate expired. With prices falling, it is worth looking at when these rates expire as we know that many of these borrowers will become forced sellers at that point.

Figure 2.8 illustrates the size of the problem. Very roughly speaking sixty billion dollars of subprime ARMs reset each quarter until 2011. The volume of Alt-A ARM resets is smaller in 2008 and 2009 but almost as big as subprime in 2010 and 2011[62]. It is likely that many of these mortgages will default unless house prices recover significantly.

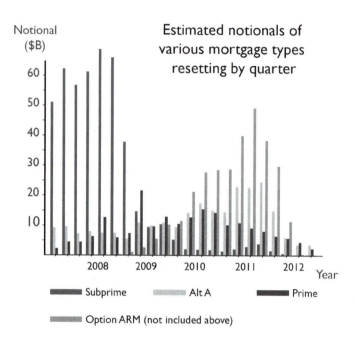

**Figure 2.8**: Estimated monthly volume of ARMs resetting

### 2.3.8 *Predatory Lending*

Borrowers and their accomplices were not the only people who were tempted to behave badly during the Boom. Lenders found various ways to enrich themselves too, some of which were either sharp practice or illegal. These come under the broad heading of *predatory lending*. The main types of behaviour here are:

- *Kickbacks*. With fees at stake for all parties, it is no surprise that relationships developed between some realtors, brokers and lenders. Whoever could extract money from the borrower passed some of that back to the others in exchange for more business.

- *Fee inflation*. Borrowers tend to shop for mortgages based on the headline rate: they often pay less attention to fees, especially when these are bundled into the loan balance. The detailed terms and conditions are sometimes not read and understood. Some lenders exploited this by charging unreasonably large fees, or by imposing huge penalties on prepayment which were not explained to the borrower.

- *Unnecessary products.* Some lenders encouraged refinancing even when it did not result in lower payments simply to generate more business. Similarly second lien mortgages or ancillary products such as mortgage insurance were sometimes sold despite not being required by the borrower or not being competitively priced.

- *Wrong products.* The fees on sophisticated loan types are usually higher than on a simple level pay structure. This encouraged some brokers or lenders to steer borrowers towards ARM products even though they may not have been the most suitable loan for them, or subprime loans even though they would have qualified for Alt-A.

### 2.3.9 Stimulating Demand

Much subprime lending was irresponsible. But was it responsible for booming house prices, or merely a mildly aggravating factor? It was certainly the main cause of house price rises in low-income areas: in 2006 over forty percent of loans in low-income census tracts were subprime[63]. Subprime lending had little directly to do with the prices of high end properties – unless they were bought by people employed in financial services – but it did drive up prices in less desirable areas.

## 2.4 A Story of the ODM: Countrywide Financial

Mortgage banks could grow rapidly when house prices were rising and there were lots of loans to be made. But pure ODM institutions were highly vulnerable to a downturn. This rollercoaster is well illustrated by the case of Countrywide Financial.

### 2.4.1 Countrywide's Growth...

Countrywide was founded in 1969 by David Loeb and Angelo Mozilo as a mortgage bank. It grew fast, as figure 2.9 illustrates[64].

The firm was on some measures the biggest lender in America during the Greenspan Boom. In 2005, for instance, it lent $494B, giving it roughly a 15% share of the total mortgage market. Countrywide was a new style institution: while it did retain some mortgages ($175B in assets is considerably more than nothing), the majority of its mortgages were sold on. Countrywide was an active lender across all segments of the market: prime, Alt-A and subprime; level pay and ARM; first and second mortgages. It lent directly, it had

|                       | 1995   | 2000    | 2005     |
|----------------------:|--------|---------|----------|
| Net Earnings          | $196M  | $374M   | $2,528M  |
| Earnings per share    | $0.49  | $0.79   | $4.11    |
| Total Assets          | $8.3B  | $23B    | $175B    |
| Share Price           | $5.41  | $12.56  | $34.19   |
| Stockholder's equity  | $1.2B  | $3.2B   | $12.8B   |

**Figure 2.9**: The growth of Countrywide Financial

relationships with brokers, and it acted as an aggregator for smaller lenders. Note, though, that growth is only possible where new mortgages are being made. Therefore Countrywide had more subprime, more speculative lending, more Alt-A, more hybrid ARMs, and more option ARMs than its older competitors. It was not just that these forms of lending were more profitable: it was also that this was where the growth was[65].

In keeping with the ODM, Countrywide kept the servicing rights for most of the loans it extended. Its mortgage servicing rights were valued at $12B in 2005, reflecting a servicing portfolio of over a trillion dollars of loans, and making it the second largest mortgage servicing provider in America. Given this growth, it is not surprising that Countrywide's principals did well. Mozilo earned over $50M in 2005 for instance.

It is unlikely that any mortgage bank would have been able to grow as fast as Countrywide did without the dramatic expansion of the Boom years. Before the Boom there were simply not enough mortgages being made to allow such a firm to increase its assets twenty-fold in ten years. More mortgages meant lower quality mortgages simply because all the good quality borrowers were already borrowing. Non-traditional borrowers did not have the money to take out conventional mortgages, so that in turn meant that the growth was in teaser rate structures, option ARMs and so on.

### 2.4.2   ...And Decline

Like so many others, Countrywide's fortunes turned with the housing market. The firm's share price started to slip in April 2007 as the market began to realise the extent of the risk posed by Countrywide's lending. Given that the firm had announced in March that 19% of its subprime loans were delinquent, it is surprising the market did not react earlier. Market participants finally understood – years too late – that falling house prices mean much higher subprime delinquencies[66] and so by September 2007, the Countrywide share price had halved.

Falling house prices mean higher Alt-A and prime delinquencies too, and this also started to become apparent in the second half of 2007[67], increasing worries about the better loans in Countrywide's portfolio.

Three things then happened at once, all bad for Countrywide.

- The market for packages of mortgage risk closed. It was essentially impossible to sell mortgages via this route as investors scrambled to reduce their mortgage exposure. This removed an important source of funds for Countrywide.

- Many mortgage lenders became distressed and a large one, American Home Mortgage, collapsed. This caused the debt markets to re-evaluate the credit quality of anyone heavily involved in the American mortgage markets, and it became much harder for these firms to borrow. As Countrywide said in a much quoted passage from its second quarter SEC filing: 'Since the Company is highly dependent on the availability of credit to finance its operations, disruptions in the debt markets or a reduction in our credit ratings, could have an adverse impact on our earnings and financial condition.' Country-wide found it much harder to fund itself, and this in turn reduced its ability to lend. The firm was forced to find new and expensive sources of funds, further damaging confidence[68].

- With falling income and rising losses, Countrywide needed more capital. It raised this from Bank of America, selling a 16% stake for $2B[69].

This kind of spiral is an inherent risk in the ODM. ODM institutions rely on being able to sell the assets they originate. If they can't, because no one wants to buy their assets, then they need to raise money some other way. But if no one wants to buy their assets then they are likely to be suffering losses and hence not be an attractive borrower. The first rule of the financial services game is 'stay in the game.' ODM institutions risk breaking that rule especially if they grow too quickly, are not well-diversified, and do not have a lot of capital.

By late 2007 there was a real risk of bankruptcy for Countrywide. It lost $600M before tax in the first nine months of 2007[70]. Various State Attorneys General had questioned its lending practices, and it was facing a number of investigations relating to predatory lending.

Despite these woes, Countrywide's origination and servicing capabilities were attractive to larger, better capitalised institutions. Bank of America bought Countrywide in January 2008, transforming itself into the largest mortgage originator and servicer in the United States. Crucially Bank of America paid for Countrywide in stock – so it did not have to raise money – and it did not immediately agree to take on all of Countrywide's liabilities. Bank of America had bought itself an origination and servicing platform without exposing itself to too many of the risks of Countrywide's portfolio. It also protected its earlier investment.

After the takeover, tellingly, the acquirer announced that they would only originate the following types of loans:

- Conforming loans (broadly well-documented loans with an LTV less than 80%);

- Mortgages with 'terms expected to produce no greater risk of default than conforming loans'; and

- Fixed rate and ARM loans subject to a ten year minimum interest only period[71].

Countrywide had left the subprime and option ARM businesses, along with most of its competitors.

Countrywide shareholders paid the price for the firm's rapid growth: the peak share price was over $45, but those that held their stock until the takeover ended up with less than a tenth of that. There was no single factor which pushed Countrywide into the hands of Bank of America. But the combination of risky lending in the retained portfolio, reliance on selling packages of loans for funding and explosive growth was too much. The Crunch had brought down the largest mortgage lender in America.

## The Dimensions of Mortgage Risk

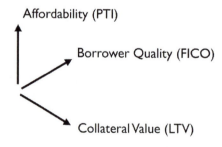

Figure 2.10: Some factors in determining the risk of a mortgage

### 2.4.3 What Went Wrong

All shareholder-owned firms – which means very nearly all firms – face the problem of balancing returns for shareholders with safety. The law obliges firms to maximise returns to shareholders. That requires growth. But when does growth endanger the firm and hence put the shareholder at risk? This question is at the heart of corporate strategy. Grow too fast and you risk blowing up, as Countrywide did. Grow too slowly and your stock price remains moribund, upsetting shareholders. In that sense capitalism is a Goldilocks business: firms need to grow not too slowly, nor too quickly, but just right.

The temptation to grow too fast, however, is much stronger than to grow too slowly. Fast growth means high profits, a fast rising share price, and big rewards for managers. The compensating control should be that the markets recognise the risk of fast growth. But in many prominent examples, Countrywide included, they haven't. Mortgage lending was a relatively low risk activity in the 1980s, with 30 year level pay structures, mostly prime borrowers, and 80% LTVs. By 2005 with hybrid and option ARMs, subprime and Alt-A, and 100% or higher LTVs[72], it wasn't. But few people noticed the change until prices started to go down.

---

# Notes

[33] For a further discussion of the particular vulnerabilities of the U.S. mortgage system, see *The housing meltdown: Why did it happen in the United States?* by Luci Ellis, BIS Working Papers No. 259 available at www.bis.org/publ/work259.pdf. The three exacerbating factors identified in that paper are a construction boom creating excess supply, laxer lending standards, and genuinely unaffordable loans (in that mortgage arrears rose *before* the economic turmoil began).

[34] For more on mortgages and the mortgage market, see *A Primer on the Mortgage Market and Mortgage Finance* by Daniel McDonald and Daniel Thornton, Federal Reserve Bank of St. Louis Review, January 2008.

[35] See *Homeowners Feel the Pinch of Lost Equity* by Peter S. Goodman, New York Times, November 8, 2007.

[36] Any excess value of the property over and above that necessary to repay the mortgage and the lender's costs typically belongs to the borrower.

[37] See www.youwalkaway.com for a website devoted to this phenomenon. In banking the term *jingle mail* refers to post arriving which contains keys. The borrower has simply sent the keys to the bank and walked away from the property.

[38] For further colour on the behaviour of some borrowers, see *Buyers' Revenge: Trash the House After Foreclosure* by Michael Phillips, Wall Street Journal 28th March 2008 available at online.wsj.com/public/article/SB120665586676569881.html?mod=blog.

[39]The 11th Federal Home Loan District Cost of Funds Index or COFI reflects the weighted average cost of funds paid by savings institutions in several Western States.

[40]This risk is sometimes mitigated by capping the maximum rate the borrower pays or by capping the maximum change from one year to the next. The cost of the cap is often met by the borrower agreeing to a minimum rate that they must pay regardless of the level of the reference index, so in interest rate derivatives terms they have bought a cap and sold a floor.

[41]Remember that a level pay mortgage repays principal over its whole life so a thirty year level pay structure is not just sensitive to thirty year rates but also to shorter term rates. See L. Hayre (Ed.), *Salomon Smith Barney Guide to Mortgage-Backed and Asset-Backed Securities* (2001) for more details.

[42]Technically a level pay mortgage is a fixed rate loan plus a strip of options giving the right to prepay the loan. The higher rate reflects the premium of those options. Hayre's book (referenced in the previous footnote) has a further discussion of the details here.

[43]For more information on interest only and option ARMs, see the Board of Governors of the Federal Reserve System, *Interest-Only Mortgage Payments and Payment-Option ARMs*, 2006. Available via www.federalreserve.gov.

[44]FICO® is a contraction of Fair Isaac. The Fair Isaac Corporation has developed retail (and other) credit scoring techniques for many years. See www.fairisaac.com for more details.

[45]This is sometimes known as a second *lien* as the lender only has a claim on the residual value of the house after the first mortgage has been repaid. For more details of equity extraction see Alan Greenspan and James Kennedy's *Sources and Uses of Equity Extracted from Homes*, Federal Reserve Board Staff Working Paper 2007-20 available at www.federalreserve.gov/pubs/feds/2007/200720/200720pap.pdf.

[46]Most HELOCs are a form of ARM based on the Prime rate, so the interest rate paid on the HELOC balance at any given point depends on the HELOC spread and the current level of Prime.

[47]The figure was 58% according to Wholesale Access. See their handout for the 2007 MBA Convention available on www.wholesaleaccess.com for more details.

[48]Even worse, the broker's fee is often not fixed, but rather includes a spread to the rate that the borrower pays. This *yield spread premium* is often not transparent to the borrower, making it hard for them to compare broked with non-broked loans.

[49]Indeed so much so that primary lenders sometimes rely on a *subservicer*: this is a specialist company that actually performs the servicing for a primary servicer.

[50]We use 'bank' to mean an institution which both takes deposits and makes loans. In most countries this is a regulated activity which requires a *banking license*.

[51]We will deal with what the broker/dealer does with the loan in detail later.

[52]For more details see *How Widespread Appraisal Fraud Puts Homeowners at Risk*, Dēmos Institute, March 2005. Available at www.demos.org/pubs/home_insecurity_v3.pdf.

[53]See www.mortgagefraudblog.com for many more examples of borrower, appraiser, broker, and other types of mortgage fraud.

[54]Suspicious activity reports relating to mortgage fraud increased by 1,411% between 1997 and 2005 according to the Financial Crimes Enforcement Network. See www.fincen.gov/news_room/rp/reports/pdf/MortgageLoanFraud.pdf.

[55]Note that some American states restrict the application of prepayment penalties making it harder to lock borrowers in.

[56]The data here is taken from testimony by Sandra Thompson before the Senate Committee on Banking, Housing and Urban Affairs, *Mortgage Market Turmoil: Causes and Consequences*, March 2007 available at `banking.senate.gov/public/_files/ACF45B6.pdf`.

[57]See the testimony of Sheila Bair before the U.S. House of Representatives Subcommittee on Financial Institutions and Consumer Credit, *Subprime and Predatory Lending*, March 2007. This is available at `www.house.gov/apps/list/hearing/financialsvcs_dem/htbair041707.pdf`.

[58]For further analysis of this, see table 1 in Fannie Mae's *Economic & Mortgage Market Developments* April 2007, available at `www.fanniemae.com/media/pdf/berson/monthly/2007/041207.pdf`.

[59]Eric Rosengren quotes an average subprime teaser rate of 8.5% in 2006, resetting after the teaser period to Libor plus 610 b.p.s in his speech *Subprime Mortgage Problems: Research, Opportunities, and Policy Considerations* available at `www.bos.frb.org/news/speeches/rosengren/2007/120307.pdf`.

[60]Various factors including loan defaults and provisions for them complicate this picture but the essential logic is correct.

[61]Strictly the present value of 3% a year for five years, but again this detail does not affect the argument.

[62]For a detailed account of the ARM reset issue see *Payment Reset: The Issue and the Impact* by Christopher Cagan, First American Core Logic Research, March 2007. This is available at `www.facorelogic.com/uploadedFiles/Newsroom/Studies_and_Briefs/Studies/20070048MortgagePaymentResetStudy_FINAL.pdf`.

[63]Data from the Joint Center for Housing Studies of Harvard University *The State of the Nation's Housing 2008* referred to previously.

[64]Much of the data comes from the SEC's Edgar database. See for instance `edgar.sec.gov/Archives/edgar/data/25191/000110465906012770/a06-2962_110k.htm` for a Countrywide SEC filing from 2005. Share prices are for the last trading day of the year, while the other data typically pertains to the firm's year end (which does not coincide with the calendar year end).

[65]For more background see *Inside the Countrywide Lending Spree*, Gretchen Morgenson, New York Times 26th August 2007.

[66]For a comprehensive discussion see *Subprime Mortgage Delinquency Rates* by Mark Doms, Fred Furlong and John Krainer, Federal Reserve Bank of San Francisco, November 2007 available at `www.frbsf.org/publications/economics/papers/2007/wp07-33bk.pdf`.

[67]See *Even best borrowers are falling behind, lender says - Countrywide's profit and stock price fall as more 'prime' mortgage holders miss payments* by Annette Haddad, Los Angeles Times, July 25th 2007, available at `articles.latimes.com/2007/jul/25/business/fi-countrywide25`.

[68]The architecture of the U.S. financial system – discussed in chapter 5 – is important here: Countrywide drew down on a previously-arranged multi-billion dollar line of credit and borrowed over fifty billion from a Government Sector Entity, the Federal Home Loan Bank of Atlanta, pledging over sixty billion of mortgages as collateral. These were the only places Countrywide could turn: given the firm's credit quality and the falling value of its assets it is unlikely that Countrywide would have found the tens of billions it needed via new private sector borrowing. (The main mortgage GSEs were of little use – Countrywide could not sell many of its mortgages to Fannie and Freddie as most of the mortgages it held were non-conforming.)

Given no one else would lend Countrywide that amount of money, the FHLB loan could be seen as implicit government support.

[69] Bank of America bought convertible preferred stock rather than simple equity, so they had the right to an equity investment in Countrywide if its stock price went up.

[70] The figure comes from Countrywide's 10-Q filing of 9th November 2007. This, and the firm's other filings, are available at `about.countrywide.com/SECFilings/SecFilings.html`.

[71] This provides borrowers with much more security of payments than a one to three year teaser rate structure: in addition, it provides Bank of America with ten years for house prices to go up. For more background see the press release *Bank of America Completes Countrywide Financial Purchase*, July 1 2008, available at `newsroom.bankofamerica.com/index.php?s=press_releases&item=8202`.

[72] The LTV can go above 100% once closing costs are built into the mortgage. In some cases the loan was larger than the value of the house even without this effect: borrowers were given a higher loan to allow them to purchase furnishings and other domestic goods.

# Chapter 3

---

# Financial Assets and Their Prices

*When the price of any commodity is neither more nor less than
what is sufficient to pay the rent of the land, the wages of the labour,
and the profits of the stock employed in raising, preparing,
and bringing it to market, according to their natural rates,
the commodity is then sold for what may be called its natural price.*
Adam Smith

---

## Introduction

The financial system distributes cash, taking it from those who have it to invest
and lending it to those that need it. At heart, there are two ways this is done.

- The *banking model* involves a bank taking deposits and using them
  to fund loans.

- The *capital markets model*, in contrast, is based on those wanting
  cash issuing *securities* – equities, bonds and such like – to investors.
  The investors pay the cash which the issuer wants: in return, they
  get whatever rights are promised in the security. These securities
  can then be traded: often they pass through many hands.

Mortgages and house prices dominated the first stage of the Crunch
as market participants digested the impact of the lending of the boom period. Attention then focussed on the holders of the risk of these mortgages:
these were mostly financial institutions. Some of the holders were active in
the banking model of finance, the capital markets model, or both. Damage
to financial institutions meant that credit became less readily available: this
slowed growth as those needing cash had to pay more to borrow. At the same,
time losses accelerated for the holders of many types of securities as the market reacted to tightening credit. Here we will examine the securities part of
the puzzle and look at how these instruments transmitted risk around the
financial system.

## 3.1   Securities

A *security*[73] is an agreement between an issuer and a holder. It entitles the holder to certain payments under certain conditions, and it may also grant other rights.

There are three very common situations.

- In a *fixed rate bond*, the payments are comprised of a fixed coupon, perhaps 3% every six months, for a fixed period. At the end of this period the *principal* of the bond is repaid. Thus in a ten year 6% semi-annual pay bond, an investor might pay £1 million initially. Every six months they receive 3% of that million, or £30,000. Then after ten years, if all goes well, the million is repaid. Note that this is in contrast to a level pay mortgage: the mortgage has a higher coupon rate but principal is repaid over the whole life of the borrowing; for the bond, it is only repaid at the end.

- In a *floating rate bond*, the interest payments vary based on some index, often Libor. Thus for instance every three months the investor might receive an interest rate equal to whatever three month Libor is, plus 1%.

- In an *equity*, any payments are entirely at the issuing company's discretion. These *dividends* are paid if the company's profits are high enough. In addition the equity may well include voting rights, such as the right to vote at elections of the company's board. Typically for an equity these rights are perpetual: the holder has the right to receive dividends and to vote for as long as they hold the security and the company still exists.

One slightly confusing piece of terminology concerns 'assets' and 'liabilities'. A security is an asset for the holder: they have the right to the benefits it conveys. On the other hand it is a liability to the issuer: they have to meet the obligation imposed by the security.

### 3.1.1   Primary and Secondary Markets

*Primary Markets* are where securities are initially sold. In these markets the issuer receives the proceeds from a sale of securities. The sale is typically managed by an investment bank[74] who may *underwrite* some or all of the issue, that is undertake to purchase the new securities if there is not sufficient interest from other buyers.

Once the securities have been sold into the primary market, they begin trading in the *secondary market*. Here holders buy and sell securities for their own benefit. The existence of the secondary markets allows the primary markets to function more efficiently: securities holders know (or at least believe) that they will be able to sell their securities if they need to; frequent issuers have clear price benchmarks for new issues; and underwriting is less risky. Often secondary market trading is intermediated by a *market maker* – a financial services firm which provides liquidity to a securities market in the hope of earning a profit.

### 3.1.2 Equity, Debt and Leverage

One key difference between equity and debt is that the obligation to pay interest on debt is usually binding: if a firm cannot meet interest or principal payments, it will be forced to default. Dividend payments are entirely optional: firms have complete discretion in paying these, and there may well be no payments if the firm makes losses.

Typically at maturity the best thing that can happen to a bond holder is that they are paid the expected amount. An equity holder has a claim on the entire value of the firm after all other claims have been met. The value of this claim – the equity price – can go up arbitrarily far if the corporation's profits increase faster than its liabilities.

Equity holders expect the most return because they take the most risk: their claim is *subordinated* and dividends are discretionary. Moreover the funding they have provided is perpetual: a share lasts as long as the company does, and is never repaid.

The optional nature of dividends and the perpetual life are what make equity loss-absorbing. Since the money raised by issuing equity never has to be repaid, a firm can survive losses provided they are no bigger than the funds belonging to shareholders. Capital or *shareholders' funds* are thus money that comes either from issuing new equity or from retaining earnings[75].

The leverage for a whole firm is defined in the same way as for a speculation: it is given by the ratio between the value of the firm's assets and its capital. Thus if we have £8 of capital and £92 of debt supporting £100 of loans, leverage is 100 : 8 or 12½ : 1. As with a speculation, the higher the leverage, the riskier the firm is, since there is less capital available to absorb losses[76].

If the firm makes a profit of £2 and retains £1 of this, then its capital rises to £9, and so based on the same leverage, it can now support £112½ of assets. Similarly if it loses £1, then its capital falls to £7, and now it can only support £87½ of assets without increasing its leverage. And if it loses £9, then it is insolvent and must declare itself bankrupt.

### 3.1.3   Bonds and Credit Spread

Bonds are issued by a wide variety of parties. Governments issue them, and bonds issued by the U.S., British and German governments are considered the safest investments in U.S. dollars, Sterling and Euros respectively. Many companies issue bonds too. Banks are active issuers, as are some industrial companies. Since companies can default, investors demand extra compensation for this risk over and above the interest that the government is prepared to pay.

Most of the time the interest rate that large banks have to pay to sell their debt is given by the *Libor curve*. Banks have to pay more to borrow money than the government so the Libor curve sits above the government yield curve. It is typically roughly the same shape, with the difference between the curves increasing slowly as the period of borrowing increases. Figure 3.1 illustrates the usual arrangement: for this curve a large bank would pay 3% to borrow for three months, rising to more than 5% to borrow for twenty years[77].

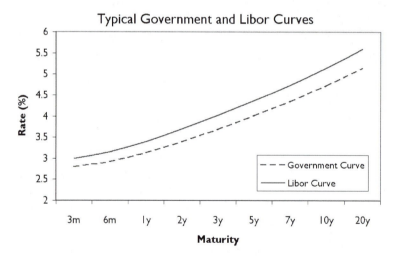

**Figure 3.1**: Government and Libor Curves

Firms with a lower credit quality have to pay more interest than large banks in order to persuade investors to buy their debt. They are said to trade at *a spread* to the Libor curve, and this spread is the difference between what they must pay and the Libor rate. Thus if ten year Libor is 5.15% and a small bank has to pay 6% on its bonds, then the bank is said to have a *credit spread* of 0.85%. Credit spreads are usually quoted in *basis points*: these are just given by multiplying the percentage spread by ten thousand, so 0.85% is 85 b.p.s.

### 3.1.4 Ratings Agencies

A ratings agency is an independent company that assesses the credit quality of debt securities and assigns a score to them based on that assessment known as a *credit rating*. We mentioned their rating of packages of mortgage risk previously, in section 1.2.4. For bonds issued by governments or by corporates, their assessment is based on (amongst other things):

- The issuer's ability and willingness to repay the debt;

- Their history of and strategy for borrowing and repayment;

- Their current leverage and the extent and volatility of their assets and liabilities;

- The consequences for the holder of the instrument of default or other failure to pay by the obligator.

Typically ratings are broadly consistent with credit spreads, so that a AAA-rated bond would have a narrow credit spread – perhaps single digit or low tens of basis points – and a CCC bond might trade at a spread of some hundreds or even thousands of basis points. Sometimes, however, the ratings agencies lag behind the markets. Thus in the Crunch we have seen the markets demanding hundreds of basis points compensation for risk the agencies still claimed was AAA. Usually the markets prevailed, and the bonds were downgraded in due course.

Figure 3.2 shows the average credit spread for bonds rated BBB (i.e. investment grade securities) and bonds rated B (junk) as a function of maturity. This is the typical situation: lower credit bonds have a higher credit spread to reflect their additional risk, and for both ratings credit spreads increase with the duration of the instrument[78].

### 3.1.5 Bankruptcy and Seniority

If a company defaults, it usually enters into the *bankruptcy* process. The precise details of the process vary from jurisdiction to jurisdiction, but typically there are two choices: either the firm is reorganised, or it is liquidated. In either case an independent party is appointed to protect the interests of the firm's creditors.

Once the bankruptcy process has been completed either through reorganisation, the sale of the firm as a going concern, or its liquidation, a certain amount of money is available to pay creditors. Imagine a queue of people waiting to be paid on a first come, first served basis. *Seniority* is the term for where in the queue you stand: this is a property of the kind of obligation

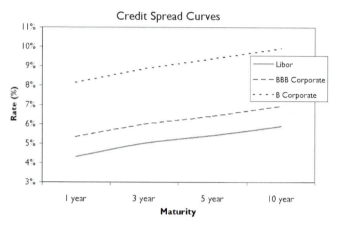

**Figure 3.2:** The term structure of average credit spreads for an investment grade and a junk rating

held. The most senior creditor presents their claim first, and is paid in full (or as fully as the funds available allow). Then the next most senior creditor is repaid, and so on. Typically most bond holders (also known as *senior* debt holders) rank at the same level in the queue: next come any subordinated creditors; and finally the equity holders.

Obviously the more senior your claim, the more chance you have of getting paid in the event of bankruptcy. Thus senior debt holders often recover some of the amount they are owed – known as the senior debt *recovery value* – whilst equity holders often get little or nothing. This is one of the main risks that equity holders take.

Note, by the way, that the terms 'stock', 'shares' and 'equity' tend to be used interchangeably: they usually refer to the least senior instrument issued by a corporation. 'Bond', in contrast, implies a more senior instrument.

### 3.1.6    Secured Bonds

So far we have discussed bonds that are issued by governments or corporates in their own name. They reflect a promise to repay by the issuer. This promise is usually just that: a promise without any further support. The bond is therefore referred to as *unsecured*.

If the markets demand too high a credit spread from an issuer for unsecured debt, there is an alternative. Just as in a mortgage, the issuer can pledge an asset as collateral against the debt. This gives rise to a *secured bond*. In the early days of the international bond markets, the commonest type used to be a bond secured against property. Thus for instance a company might pledge its head office building as collateral against a bond. This would make the bond safer and allow the company to pay a lower credit spread. (On the other hand, of course, this reduces the assets available to unsecured creditors in the event of bankruptcy. Logically then when a company issues a secured bond, the credit spread of its unsecured debt should increase.)

The next logical extension of secured debt is to take the company out of the picture entirely. Consider a pool of mortgages. (Most of) these will continue to pay regardless of whether the originator is still around or not. Therefore if we can use the cashflows from the mortgages to repay debt, that debt might be better credit quality than the originator. In particular we need to give the bond holders the right to receive any and all payments from the mortgages regardless of the health of the originator. This leads us to the idea of an *asset backed security*.

### 3.1.7 Asset Backed Securities

Suppose a firm has a collection of assets, perhaps because it has originated them. This may be a portfolio of mortgages, corporate loans, or even something like the right to receive royalty payments on David Bowie's back catalogue of recordings. What is important is that the assets pay cash regularly and reasonably predictably. The owner can raise money by using their assets to back the issuance of a security. Here's roughly how it works.

The firm sets up a company known as a *special purpose vehicle* or SPV. The SPV buys the assets. Thus for instance a mortgage bank might set up a SPV which purchases some of the mortgages it has made. Suppose the SPV buys $1B of mortgages. It gets the money to pay for the mortgages by issuing bonds to outside investors. Investors are comfortable buying the bonds because the SPV issues fewer bonds than it has mortgages: the mortgage bank's SPV might only issue $950M of bonds for instance. The other $50M comes from less senior securities: the SPV issues shares, some of which are bought by the mortgage bank, and some by other parties.

The bonds are known as *asset-backed securities* or ABS since the only thing protecting the investors are the underlying assets: they will only get their $950M back if total losses on the mortgage pool do not exceed $50M.

Notice that if losses on the mortgages are only $30M, the bonds will be repaid, and the share of the SPV will be worth $20M. Because a 5% loss rate on prime mortgages is (absent a deep recession) unlikely, the asset backed securities are highly rated and so investors are willing to buy them even though their credit spread may not be very high. In particular the interest rate paid on

these bonds will be less than that paid by the mortgages, so there will probably be *residual* cash left over to pay a dividend to the equity holders. This
situation is illustrated in figure 3.3.

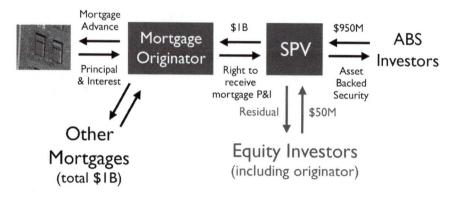

**Figure 3.3:** Outline of a simple ABS issuance structure

The mortgage bank likes the structure because it gets cheap funding for
95% of its mortgages, and it still has a stake in the performance of the pool via
the equity it retains in the SPV. Even better, by setting up a separate company,
the originator does not have to count the SPV's debt amongst its liabilities;
nor does it have to count the SPV's assets as its assets. If the originator is
regulated it does not have to keep capital against the assets either[79].

## 3.2    Markets and Prices

A market is nothing but a place where sellers and buyers meet. The price of an
asset is just the level at which a willing buyer and a willing seller are prepared
to transact. Despite this, there is a temptation to imbue market prices with
a greater meaning than this, perhaps suggesting that they somehow represent
a 'fair' value for the things which are bought and sold. When might this be
reasonable?

### 3.2.1   *What Are Prices Good For?*

If we wish to value an asset, one approach is to estimate the price at which
we could trade it[80]. This route tends to be more reliable when:

1.    There are many assets which are the same, or very similar, and these
assets are well understood by market participants;

2. There are many independent buyers and sellers of this collection of assets, and they trade frequently;

3. The buyers' and sellers' ability to trade does not depend strongly on factors beyond their control;

4. A large amount of trading happens in the market, and every buyer and seller can readily discover the price at which recent transactions have happened;

5. The market has high standards of conduct which are actively policed. In particular there are rules against market manipulation.

Suppose you own a painting by Velázquez. There is no daily market in the thing you own. Even comparable assets – other paintings by Velázquez – trade infrequently. If you want to sell your painting, you will have to arrange an auction, hold viewings for interested parties, subject the painting to authentication, and so on. It could take months, and the result is uncertain. The final price will probably be a result of a tussle between a few major museums and the wealthiest private collectors. A 'fair' price does not make sense except in comparison with the prices of other assets, and any given Velázquez is literally incomparable to anything else. The first condition above ensures that we are not in this situation, but rather dealing with the price of well-known, often-traded assets.

The second condition gives us depth to the market: it means that many transactions will happen, and that even if we have a significant volume to buy or sell, the impact on the price will not be too severe. Moreover, if the third also holds, it means that these buyers and sellers are unlikely to disappear because of some external event.

Price transparency is important because it means that we have information about what an asset is worth today without having to test the market: we can use market prices to value our position. Finally good market regulation makes it more likely that we can take market prices at face value.

### 3.2.2 Trading Securities

Securities trade by a variety of means. These include stock exchanges (which confusingly sometimes trade bonds as well as stocks), electronic exchanges, and market makers who buy and sell securities directly. In many cases this trading is reported, so we do indeed have some measure of price transparency for many securities.

The principle of valuing financial instruments using current market prices is known as *fair value*. For many securities, this process is uncontroversial:

one simply looks at the market prices prevailing currently. However it is by no means the case that all securities trade regularly, and some financial assets – such as many loans – do not trade at all. Thus in order to understand valuation, we need to look at the liquidity of financial assets.

## 3.3    The Liquidity of Financial Assets

Financial assets are *claims*: they convey rights, or promises of later payment. The two models of financial services activity – the banking model and the capital markets model – generate different assets. Most of the assets of an old-style bank are loans, whereas a capital markets participant will typically have more equities, bonds, or both. Clearly the asset quality of a firm has a strong impact on the firm's riskiness: good quality assets are unlikely to imperil an institution. But also important is the *liquidity* of the firm's assets: if there is a problem, how quickly can the firm sell its holdings? How much are they likely to get if they do? How much might selling change market prices? We begin by looking at these dimensions of *asset liquidity risk*[81].

### 3.3.1    What is Asset Liquidity?

Liquidity risk is interesting – and complicated – because the questions above cannot be answered precisely. For some assets, such as a hundred thousand pounds worth of a major stock, we can be fairly sure that we can sell whenever the market is open, and fairly sure that the price we will get is close to the published price. A trader can look on a screen, see the price, and *know that that really is the price at which the asset can be liquidated*. It makes sense to call this kind of asset *liquid*.

Other assets are definitely illiquid. Exactly how much you get for a Velázquez will depend on who wants the picture the most and how much they can afford to spend. Unlike shares in well-traded corporations, then, old master paintings are not a great asset for an institution that needs to raise funds *right now*.

All of this would be fine if the liquidity characteristics of an asset were fixed. When we buy a painting, we do not expect to have to sell it in a hurry. Some securities, however, are liquid for a while, but then turn illiquid. This is much more dangerous. We think we have something that we can liquidate in a hurry if we need to, but then the markets turn against the kind of security we have, and suddenly we are stuck with it for months or years.

### 3.3.2 Loan Liquidity

Most single loans are illiquid. Unless the borrower is a well-known company, a loan cannot easily be sold. If a firm wishes to liquidate a loan early – before it has been repaid – then the best approach is often to package it up with a number of other loans and try to sell the portfolio. (This process of *securitisation* is discussed in detail in chapter 6.)

### 3.3.3 Security Liquidity

Security liquidity varies dramatically both between securities and over time. Government bonds from the largest issuing countries are usually very liquid. Some bonds issued by corporations are also liquid, especially when the bond issue is large and the corporate is a frequent issuer. But there are also many smaller, less liquid bond issues. Some bonds are sold once and never trade again: this is particularly the case for unusual, structured instruments.

The equities comprising major indices such as the S&P 500 or the FTSE-100 are typically highly liquid. However very large equity positions or positions in smaller companies can be less easy to sell.

### 3.3.4 Equity Market Liquidity

The reason that we were reasonably upbeat about the liquidity of major stocks was that equity markets – mostly at least – fulfil the criteria of subsection 3.2. All equities are the same, so the price of a trade in a thousand shares of a corporation is a guide to the price we could achieve if we held ten thousand shares. Moreover prices on most exchanges are *reported* in a timely manner, so we have information on actual transactions. Finally there are a diversity of market participants including retail investors, pension funds, hedge funds, and investment banks, so liquidity droughts caused by events like the Crunch tend not to be too bad.

### 3.3.5 Bond Market Liquidity

Some bond markets are also liquid. But the markets for many corporate bonds and for ABS are not. There are a much smaller number of market participants here than in the equity market, and many of them are leveraged institutions. Therefore in a liquidity drought – when borrowing to invest in financial assets is difficult and financing is expensive – these institutions tend to leave the market.

Even worse, at least for ABS, the pool of identical assets is small. Major corporations have tens or even hundreds of billions of dollars of equity in issuance, but an ABS deal may only be a few hundreds of millions of dollars. Moreover I cannot immediately compare the price of an ABS issued by one ODM bank with that of another since the asset pools are different. Bank A's mortgages may be mostly made in Florida in 2006, and thus rather risky: Bank B's were made in Manhattan and hence less so.

All of this means that the last reported price of an ABS is often a less clear guide to the value of the security than a major equity's price. There may not have been recent trades, or the trades that have happened have been forced sales. The prices of other ABS are useful information but not a definitive guide. There is uncertainty in the value of the instrument.

## 3.4  What's In It For Me?

In this section, we look at the profit available to a financial institution from making a loan or buying a bond. This profit, and the factors which can affect it, are the key to understanding which assets firms buy or sell.

### 3.4.1  Loans in an Old-style Bank

Consider a large bank which funds at Libor flat and suppose the bank makes a loan at a credit spread of 85 b.p.s. This gives 85 basis points of profit per year. What is this profit for? An old-style bank is being compensated for (at least) two risks in making a loan:

- Firstly and most importantly the loan might default. The bank needs to be compensated for this *default risk*, and some of the credit spread is for that[82].

- Regulators require that a bank puts up capital against its loans. This capital has to earn a return. Therefore another thing the bank is receiving compensation for is the cost of the capital required to support the loan.

### 3.4.2  Bonds Are Different

Now consider a firm that buys a bond. In some ways this is similar to making a loan: the firm has the risk that the bond will default, just as a lender has default risk on loan counterparties. The bond pays principal and interest, just like the loan, so we can talk about its credit spread (i.e. how many basis points more than Libor we receive in interest if we buy it). But once we think about a firm holding a bond rather than an old-style bank making a loan, that credit spread is compensation for more than just the risks we discussed in the last section.

One reason for this is that many bonds are *marked to market*. That is, every day the holder values the bond. If it has gone up since yesterday, they record a profit. If it has gone down, they take a loss. This daily mark to market is possible since bonds – some bonds anyway – are liquid. They have market prices. Loans don't, so we cannot mark them to market. Mark to market is desirable when market prices give the best view of what an asset is worth.

The key difference between how loans are accounted for and fair value securities is that loan accounting takes no regard of any current information. Suppose my company takes out a £1 million loan for ten years at an interest rate of six percent. After that, sadly, my firm's credit rating is cut twice and persistent rumours go around the market about its solvency. That has no accounting impact for the bank I have borrowed from. As long as I keep paying interest, they recognise the loan as an asset worth £1 million. They make the assumption I am going to pay them back[83]. On the other hand, if I had issued a bond and the purchaser was marking to market, they would suffer losses as the market price of my debt declined, reflecting the bad news about my credit worthiness. This loss is *unrealised*: it has not happened yet, but it is reflected in the earnings of a firm using fair value accounting.

### 3.4.3    What is the Credit Spread of a Bond For?

Here are some of the non-default risks that a mark-to-market bond investor might require compensation for[84]:

- *Funding risk.* Many holders of securities have to *fund* them. If a bond can always be funded easily[85] then this is not a concern. Securities whose funding is costly or uncertain pose an additional funding risk to the holder which may require compensation.

- *The volatility of the bond price.* As the bond price changes, marking to market causes gains or losses. Losses have to be supported by capital, since without sufficient capital to absorb losses, firms become insolvent. Thus the higher the volatility of a bond's price, the more capital is needed to support possible losses. This capital has to earn a return, so higher volatility securities need to offer their holders bigger spreads.

- *Current liquidation risk.* If an asset is known to be illiquid, then we can assume it will take some time to sell. During this time the asset may fall in value. Therefore current fair values may not reflect actual future liquidation values. This risk may again require compensation[86].

- *Uncertainty in future liquidation risk.* Even if an asset is currently liquid and can be sold with little or no delay, it may become illiquid in the future. Asset holders may demand compensation for this potential future illiquidity[87].

- *Uncertainty in current valuation.* If the fair value of an bond is uncertain, then it is prudent to take a sum of money aside to cover this uncertainty. Thus if I have a bond whose fair value is between $1M and $1.2M but I do not know which, perhaps because it has not traded for a little while, then it would be imprudent just to mark it at the best guess, $1.1M (and even worse to mark it at $1.2M). Good practice would be to use the best guess, $1.1M, but then to take an extra $0.1M aside as a *valuation adjustment*[88]. This adjustment reduces the profits of the institution and hence firms may demand an extra return from the bond to compensate them for the funds tied up in the valuation adjustment.

- *Uncertainty in future valuation.* Similarly if future valuation may be uncertain – due to potential future illiquidity – firms may demand compensation for potential future valuation adjustments.

We will call any compensation a bond holder receives above and beyond fair compensation for default risk a *non-default component of the credit spread.*

### 3.4.4 *Why Credit Spreads Matter*

Before the advent of the ODM, most retail mortgages were held in old-style banks. These banks only need sufficient returns to compensate them for the default risk of their mortgages, plus some profit on the equity supporting the book. They are indifferent to the market prices of mortgage assets. All that matters is the *realised losses* on the mortgages: how many of them are delinquent; how many of them default.[89].

Once the ODM became established, however, this changed. ODM banks packaged up their mortgages into ABS. Many of these bonds were bought by investors who marked them to market. Now the market price of these bonds matters too. If they fall, the holders take losses. These losses erode capital, regardless of the long term outlook for the mortgages. The bond holders may decide to sell even though they believe that the long term outlook is good. In short, fair value gives extra visibility of the risks an institution is taking, and often it results in increased earnings volatility – something the firm's managers may find unwelcome[90].

Note that the market can perfectly rationally bid down the prices of bonds *even if it does not believe that the risk of default has increased.* If funding costs go up or liquidity goes down, bond buyers will need more return to compensate them for these risks. Therefore they will pay less for bonds.

Thus in the Crunch we found the holders of mortgage assets, troubled by increasing funding costs and decreased liquidity, selling while they could. This of course caused the price to drop further and these assets became even less liquid as many holders rushed to sell at once. Markets in ABS cut two ways: visible prices gave market players better risk information; but selling resulting from their risk analysis drove prices down further as buyers disappeared.

### 3.4.5 *The Effect of ABS Issuance in the ODM*

At their simplest, ABS simply pass on risk and return from one party to another in exchange for cash. Like any bond, the buyer pays cash for a set of promises and takes the risk that the promises will not be fulfilled. However there are other features of this process, too, as we have seen in previous sections:

- Before the advent of the ODM, most risk originated by banks was not marked to market. Many of the holders of ABS mark them to market. Therefore there has been an *accounting transformation.*

- Because ABS sometimes trade, market values are sometimes available. This means that there is much greater *visibility of the price of risk* than before. That price may not reflect just the market's expectation of the fundamental long term performance of the ABS, but not all market participants realise this.

- When risk was held in an old-style bank, the risk holder was also the originator. All of the information that was available at origination was in principle also available for risk management purposes. However once risk is passed on via ABS, that information may not be available to the buyer. *Information asymmetries* can develop whereby the buyers of risk know less than the sellers.

- Similarly *servicing risk* arises when the servicer is not the risk holder. They may not be incentivised to do their job as well as the risk holder would want them to.

- Lastly, no matter how legally robust the securitisation process is, the originator may still feel a certain responsibility towards the performance of ABS backed by assets they have originated. There may, in other words, be *reputational risk.*

### 3.4.6   What Prices Mean

A market price is not a fundamental assessment of value. When there are prices – that is, when there are sufficient buyers and sellers for a price to be discoverable – these prices simply reflect where these parties are willing to trade. The decision to trade is based on a variety of factors including funding costs, liquidity, and price volatility, as well as expectations of long term fundamental value. Losses in fair value can therefore occur not just because perceptions of fundamentals deteriorate, but also because funding costs increase or liquidity decreases.

Thus even if you believe that prices are based on perfectly rational assessments, they include more than the impact of estimates of fundamentals.

Fundamental value is only known after the fact. The behaviour of a typical mortgage is only certain after thirty years. Moreover even if we could strip out expectations of fundamentals from other market factors, there is no reason to believe that the average of these expectations will accurately predict the future[91].

There is another factor too which makes market prices even less reliable guides to fundamental value: that is other people[92]. Asset buyers sometimes buy because they think that the prices are going up, or more precisely because they think that the total return from an asset will be more than commensurate with the risk that they run in holding it, given their costs. Asset returns depend on the actions of other market participants: if lots of them buy, prices go up. Fundamentals in this reading are irrelevant[93]. Rather, it is your expectations of what other people will do that determines your investment decisions. This phenomenon – sometimes called *reflexivity* – means that prices can change rather quickly if the general view of the market changes. When there is confidence, people expect other people to buy. So they buy too, and prices go up. Actions validate beliefs. But when confidence falls, selling is expected, so it happens. The market is vulnerable to a paradigm shift[94].

### 3.4.7   The Variation of Asset Returns

Whatever the causes of price movements, asset returns do sometimes show larger swings than would be expected simply based on random variations. For instance figure 3.4 illustrates the number of large movements on an equity index actually experienced, and the number predicted based on independent random movements. The model used for the predictions – known as the *normal returns model* – is clearly inadequate to describe the actually observed frequency of big falls on this index. Big falls happen far more often than

the idea that the normal returns model would suggest. This phenomenon –
known as *fat tails* – is widespread[95].

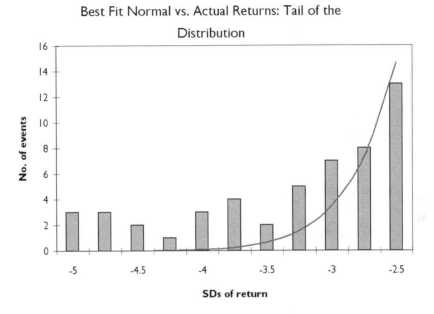

**Best Fit Normal vs. Actual Returns: Tail of the
Distribution**

Figure 3.4: Fat tails in an equity index

Finally of course there is an implicit assumption in the analysis of prices
that market participants act rationally given what they know and how they
are compensated (or in the reflexive reading, what they know, how they are
compensated and what they believe about the markets and other market participants). We will not dwell on the extensive literature which questions this
assumption, but it is at least worth noting that it is not correct to assume
rationality in many situations[96].

## 3.4.8   *The Interaction between Rates, Asset Prices, and Beliefs*

One concrete mechanism by which beliefs influence asset prices comes in
periods of lower interest rates. When interest rates are low, companies find it
cheaper and easier to borrow, and so they invest more in their business. Fundamentals improve, increasing asset prices. The general sense of confidence
generated by cheap money and rising asset prices encourages further investment. So far we have a virtuous circle. But it turns nasty: leverage is cheap, so
speculation increases. If interest rates remain low, asset price bubbles can develop. Moreover if financial asset prices rise too quickly, speculation becomes

rational compared with investment in real assets or R&D, so corporates become speculators too. All that is necessary for the spiral to continue is for most investors to believe that more money will be spent on financial assets in the near future, and for there to be no intervention by external forces. If beliefs change or externalities intrude then the bubble will burst and asset prices will fall fast.

### 3.4.9   Pre-Crunch Conditions

The Greenspan Boom offered ideal conditions for an asset price bubble to develop. Rates were low, confidence was high, and rising growth generated more and more investable funds. These funds were naturally attracted into financial assets, and there were few regulatory or other obstacles to prevent rapid growth in financial asset prices[97].

| Credit Spread in Basis Points | | |
|---|---|---|
| Date | Good quality U.S. bonds | U.S. junk bonds |
| October 2002 | 131 | 786 |
| February 2003 | 81 | 662 |
| June 2003 | 62 | 553 |
| October 2003 | 39 | 419 |
| February 2004 | 35 | 358 |
| June 2004 | 29 | 323 |
| October 2004 | 31 | 308 |
| February 2005 | 28 | 257 |
| June 2005 | 37 | 318 |
| October 2005 | 28 | 303 |
| February 2006 | 32 | 268 |
| June 2006 | 29 | 254 |

**Figure 3.5:** The collapse of corporate credit spreads during the Greenspan Boom years

From late 2002 to 2006 credit spreads on corporate bonds and ABS reduced dramatically, as illustrated in figure 3.5. Large amounts of money were invested in the bond markets. Falling spreads meant that, in order to get a reasonable return, investors had to take more and more risk. For instance the figure shows the compensation for holding good quality bonds decreased by a factor of approximately 4, whereas it was about 3 for junk bonds. This collapse in credit spreads produced a *Hunt for Yield*.

By 2006 it had stimulated the creation of a wide range of products aimed at increasing investor returns, some of dubious sanity[98], and very nearly all involving extra risk.

There were, of course, many commentators who warned of the impending problems presaged by tight credit spreads (although perhaps not quite so many as claimed that honour subsequently). But in any event predicting a crash is vacuous: there will *always* be a crash sometime in the future. No firm will desist from participating in the market simply because disaster could come at any moment, because they know that their peers will persevere, and failing to participate will reduce profits[99]. Thus market participants ran headlong into the Crunch. Some of them would not survive: others would only make it thanks to considerable state aid. Much of this aid would be provided by Central Banks, so in the next chapter we turn to these institutions, their rôle in ordinary conditions, and how they tried to help financial institutions when the Boom ended.

---

## Notes

[73]This section is based on material from Chapter 1 of my book *Understanding Risk*, Chapman Hall 2008.

[74]Just as with a mortgage bank, and equally confusingly, an investment bank may not actually be a bank: they might instead be a broker/dealer.

[75]We have skimmed over some of the issues here. Capital is divided into a number of tiers. The best quality capital – Tier 1 – is share capital plus retained earnings. It also sometimes includes a small amount of *innovative Tier 1 instruments* which are somewhat like equity. Tier 2 includes less equity like instruments such as (certain kinds of) subordinated debt.

[76]There are a variety of related definitions of leverage: some compare shareholders' funds with assets; some with debt. Some definitions subtract intangible assets; others include off balance sheet liabilities. The key idea is always to estimate how big a margin of safety there is, that is how much the firm's assets can fall without endangering the enterprise.

[77]We have used the term 'Libor curve' for what some people call the swaps curve. No bank funds exactly at Libor flat all the time, but in normal conditions the Libor curve is a reasonable approximation to the cost of funds for the best credit quality banks.

[78]The credit spreads shown are based on data in Jeffrey Amato and Maurizio Luisi's *Macro Factors in the Term Structure of Credit Spreads* (2005) available at www.ecb.int/events/pdf/conferences/ECB-BIS_2005/credit-macro20.pdf.

[79]Although there will be a capital charge against the stake it holds in the SPV.

[80]Note the hypothetical here: often we are not *actually* going to trade. This subtlety has attracted some attention from those interested in the epistemology of finance: see for instance

*Accounting - Truth, Lies, or 'Bullshit?': A Philosophical Investigation* by Norman Macintosh, 2007, available at `/papers.ssrn.com/sol3/papers.cfm?abstract_id=1006875`.

[81] For a further discussion, see Avinash Persaud's *Liquidity Black Holes* (2003).

[82] Some traditional models of credit risk say that is the *only* risk that a credit spread is compensation for. For a good account of the modern version of the theory see P. Schönbucher's *Credit Derivatives Pricing Models: Model, Pricing and Implementation* (2003).

[83] My bank *might* take a loan loss provision if my credit quality declines sufficiently. But they do not have to.

[84] Other accounts of the components of credit spreads include Jeffery Amato and Eli Remolona's *The credit spread puzzle* (BIS Quarterly Review December 2003) available at `www.bis.org/publ/qtrpdf/r_qt0312e.pdf` and Ann Marie Hibbert et al.'s *Determinants of the Credit Spread: Further Evidence on the Relationship between Market and Credit Risk* (AFFI 2007) available at `affi2007.u-bordeaux4.fr/Actes/217.pdf`. Also worthy of note is Alessio Brown and Žiga Žarnić's *Explaining the Increased German Credit Spread: The Role of Supply Factors* (Kiel Institute for World Economics Working Paper No. 412) available at `www.econ.kuleuven.be/public/ndcalc9/Brown_Zarnic_ASP_WP.pdf` which makes the important point that if there are a lot of bonds being sold, they are likely to be cheaper. In normal market conditions these supply/demand constraints might not be important, but they can dominate in a crisis.

[85] We discuss bond funding further in the next chapter. The key idea is that some bonds can be used as collateral for secured funding in a *repo*. If we are sure that a bond can be funded cheaply by this means, then a lower funding premium is required than if this is not the case.

[86] Liquidity premiums – where liquid assets offer a lower return than similar but illiquid ones – have often been observed in financial assets: see F. de Jong and J. Driessen's *Liquidity Risk Premia in Corporate Bond and Equity Markets* (2005), available at `www1.fee.uva.nl/fm/PAPERS/FdeJong/corpliquid1104.pdf` and *The Flight-To-Liquidity Premium in U.S. Treasury Bond Prices* by F. Longstaff, *Journal of Business* Volume 77, 2002 available at `fmg.lse.ac.uk/upload_file/150_Francis_Longstaff.pdf`.

[87] This is particularly an issue if capital requirements depend on liquidity, as in parts of the SEC's capital regime for broker/dealers.

[88] Some firms would simply mark at \$1M, but this approach can result in assets being marked down too far without any visibility as to what is happening. The author prefers marking at the current 'best guess' then taking valuation adjustments for any uncertainty, as this makes clear the size of that uncertainty, information that is useful in understanding the risk of the firm's assets.

[89] What I am suggesting here is that *just* changing accounting from accrual to fair value changes the return a holder needs from an asset. In accrual accounting they just need to earn enough to cover default risk plus earnings on the equity support: in fair value, they need more, as we saw in the previous section. Moreover for holders subject to regulatory capital requirements, this is especially true since losses in fair value are typically deductions from capital. Thus if a security is bought at 100, falls to 98, then rises back to 100 some time later, the 2 of loss in the interim reduces the institution's capital. This institution therefore has to hold extra capital against the volatility of the fair value of its portfolio regardless of the long run portfolio return. All of these effects increase the hurdle rate of return on an asset if it moves from a setting where it is accrual accounted to one where it is marked to market.

[90] Visible prices thus couple the financial system together: a single transaction, when observed, causes all holders of the asset who use fair value to mark their books at that price.

[91] The fundamental value of a risky financial instrument can never be known: that is what we mean by it being risky. Under a collection of assumptions (which are usually of questionable veracity) we may be able to estimate that value, but it cannot be precisely known. Financial instruments are like fireworks: we only really know what they will do after they have done it.

[92] George Soros, in *The New Paradigm for Financial Markets*, 2008, makes this point forcefully, and others have commented on it too. See for instance Donald Mackie's *The Big Bad Wolf and the Rational Market: Portfolio Insurance, the 1987 Crash and the Performativity of Economics*, available via www.gloriamundi.org/detailpopup.asp?ID=453057236.

[93] It is worth noting that the reflexive view, although it has been expounded for some years and has some obvious attractions, is still not widely accepted. What is broadly accepted is that the markets are imperfect. For a good summary of the various views on credit market failings, see *Aggregate Implications of Credit Market Imperfections* by Kiminori Matsuyama, available at faculty.wcas.northwestern.edu/~kmatsu/Matsuyama\percent20NBER%20Macro%20Annual,%20Final%20Version.pdf.

[94] Thus for instance some asset prices have long periods of stability followed by short periods of rapid change. One explanation for these fat tails is that there is a shift in the beliefs of market participants.

[95] There are a number of better approaches to modelling financial return distributions than the simple normal or log normal model. For instance, Manfred Gilli and Evis Këllezi's *An Application of Extreme Value Theory for Measuring Financial Risk* (Computational Economics Volume 27 Number 1, 2006) and the references therein contain more information on extreme value theory, Thomas Lux's *The Multi-Fractal Model of Asset Returns* (Christian-Albrechts-Universität Kiel Economics Working Paper 2003-13) and its references discusses fractal approaches, and the articles in Andersen et al.'s *Handbook of Financial Time Series* (2008) contain a variety of other approaches.

[96] See *Inefficient Markets: An Introduction to Behavioural Finance* by Andrei Shleifer (2000) or *Behavioural Finance: Insights into Irrational Minds and Markets* by James Montier (2002) for more details.

[97] Soros calls the period beginning with the de-regulation of the Reagan and Thatcher years *market fundamentalism*. Certainly by 2000 the belief that markets performed best when they were interfered with least was widespread.

[98] We look at these in chapter 8.

[99] Chuck Prince, ex CEO of Citigroup, made a widely derided remark in mid 2007, saying that Citigroup was 'still dancing' (Financial Times, 9th July 2007). In a sense though Prince was right: had the markets not collapsed, his shareholders would have been furious that Citigroup had withdrawn from a profitable area when its peers were still trading. Moreover a player as large as Citigroup scaling back its activities could in itself have caused a market collapse.

# Chapter 4

---

## Liquidity and Central Banks

> *What is wanted and what is necessary to stop a panic*
> *is to diffuse the impression that though money may be dear,*
> *still money is to be had. If people could be really convinced*
> *that they could have money if they wait a day or two,*
> *and that utter ruin is not coming, most likely*
> *they would cease to run in such a mad way for money.*
> Walter Bagehot

---

## Introduction

If you had to summarise the sources of financial risk in a single word, one reasonable choice would be 'leverage'. Without borrowing, speculation might damage speculators, but it is unlikely to damage anyone else. Once we allow people to borrow, however, the whole financial system becomes risky. Failed speculators can bring down the firms who lent to them as well as themselves. Buying things on the 'never never', as my grandmother called it, is a business that needs to be carefully managed by the borrower and the lender. In particular there is an interaction between *solvency* – whether a firm's assets are worth more than its liabilities – and *funding liquidity* – whether a firm can borrow.

In this chapter we look at where that borrowed money comes from. Specifically the sources of funding for financial institutions are discussed together with the risks inherent in this process. Difficulties in obtaining funding were important in the demise of both Bear Stearns and Northern Rock so we need to understand this process in order to have a good sense of how financial institutions fail.

Central Banks have a big influence here, first through their management of the supply of money in an economy, and second since they sometimes act as *lenders of last resort* to firms. Their rôle in providing liquidity and in assuring the funding of assets was key to avoiding widespread bank failures during 2007 and 2008. Central Banks more or less made up their policies as they went along. They introduced a series of new facilities in an attempt to keep the banking system functioning as one problem after another arose. The varying success of their innovations gives us some insight into the policy instruments available to Central Bankers during a crisis.

Not everyone survived the liquidity squeeze of 2007. In particular, some ODM firms that relied on selling ABS to raise money suffered distress. We look at one of these, Northern Rock, and show how its funding strategy led to disaster.

## 4.1   The Basis of Old-Style Banking

To buy things you need money. This applies as much to financial institutions as to school children in a sweet shop. The process of ensuring that a firm has enough cash to meet its needs is known as *funding*. Mostly this process is entirely routine: raising cash is not a problem for most financial institutions, most of the time. However if cash cannot be raised promptly then the firm may fail. How do firms think about this risk?

### 4.1.1   *Where Does the Money Come From?*

Simple companies have two types of funding: *equity* and *debt*. Equity funding or *capital* comes from issuing shares or retaining profits. (These profits would otherwise be distributed to shareholders). As we discussed in the last chapter, equity funding provides the most flexibility because it never has to be paid back. On the other hand, it is the most expensive, as equity holders demand a high return for the risks they take. Whenever equity is needed, then, firms need to ensure that they earn sufficient return on it. Indeed – at least in the Anglo-Saxon model of capitalism – this is their primary duty.

The other kind of funding comes from borrowing money, either by taking deposits, borrowing from other institutions, or by issuing various kinds of bonds. This kind of funding (nearly always) requires timely payments: you have to pay interest and repay principal on your debts when you said you will or you are in default. Unlike a mortgage holder – who typically gets thirty days grace before they are called delinquent – companies have to pay their liabilities promptly.

Equity and debt funding together provide the cash that allows a bank to acquire assets like loans or securities. Investors are provided with a snapshot of the firm regularly in its *financial reporting*. One of the pieces of information given here is the firm's balance sheet: this details its assets, liabilities, and capital. The balance sheet always balances, so the size of the asset side of the balance sheet equals that of the liability side as illustrated in figure 4.1[100].

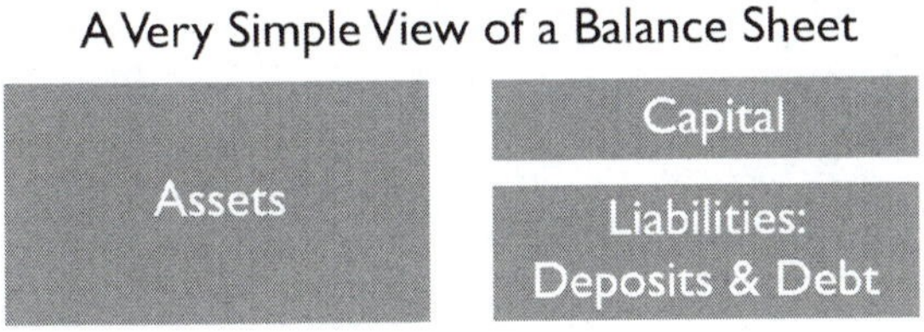

## A Very Simple View of a Balance Sheet

Figure 4.1: The top-level structure of a bank balance sheet

### 4.1.2 The Term Structure of Various Yield Curves

Typically the government and Libor curves point up, as illustrated in figure 3.1. That means that borrowing for longer periods is usually more expensive than borrowing for shorter periods. This is especially so for smaller or riskier firms as credit spreads tend to increase with maturity too. Thus for instance a big broker/dealer might fund at Libor plus 10 b.p.s for three months but at Libor plus 50 b.p.s. if it wanted to borrow money for five years.

Therefore there is a great temptation to fund long term assets with short term liabilities since it is cheaper. This gives rise to a risk known as *funding liquidity risk*.

### 4.1.3 Funding Liquidity Risk

Suppose a firm originates an asset with a ten year life – a ten year loan for £1M, say. In order to make that loan it needs to find the money. Therefore the firm borrows £900,000, with the rest coming from the firm's capital. How long should it borrow for? Clearly the safest answer is ten years: if it borrows for a shorter time, say a year, then it will have to replace that funding after the year has passed. If it cannot *roll* its funding – perhaps because the market has lost confidence in the firm and is unwilling to lend it money – then it will have to default. This is because it has taken the original £900,000 and given it to the loan counterparty. They will not repay for another nine years. So the only way to repay the one year loan is to take out another loan.

Thus the safest way to fund an asset is to borrow to *term*. If we fund a ten year loan by borrowing for ten years, we do not have to roll the funding. Many non-financial companies follow this strict discipline of funding assets to term: if they acquire a factory with an expected life of thirty years, they issue a thirty year bond to fund it. Financial institutions, however, typically don't. There are two reasons for this, neither of them particularly good.

- Matching term is expensive because borrowing for the long term is often more expensive than borrowing for the short term.

- Many financial assets are liquid. This means that if a firm has trouble rolling funding it can raise money by selling assets.

Therefore partly to make more money, and partly because some of its assets are marketable, financial firms typically do not match the term of funding with the term of their assets. They lend long and borrow short, otherwise known as taking *funding liquidity risk*[101].

### 4.1.4   Liabilities and Confidence

Funding sensibly involves keeping a balance between assets and how they are funded. This is difficult because both parts are uncertain. Let us look at the liability side first. The primary sources of debt funding for financial institutions are:

- Deposits, if the firm is a bank;

- The *interbank* and *money markets*, where short term funds are borrowed and lent between professionals; and

- The issuance of various kinds of bond.

These vary enormously in their characteristics. For instance, retail deposits are often *demand* instruments: the depositor can demand their money back at any time. But typically they don't: many people leave some cash in their deposit accounts for long periods. If all of a bank's depositors tried to take all of their money out at once, the bank would almost certainly fail. This is sometimes known as a *bank run*. It illustrates the key rôle of *confidence*: providing people do not think that a bank is likely to fail, they are willing to give it their money. And providing a bank can attract funding – either by taking deposits or by borrowing – it is much less likely to fail. Banking, then, is a kind of confidence trick. If people have confidence in a financial institution, then that confidence is likely to be well-placed. But if they lose confidence, that loss in itself makes a problem much more likely.

Retail bank runs have been unusual since the widespread adoption of deposit protection. But not all of a financial institution's liabilities are guaranteed by the state: in particular its interbank and capital markets borrowings aren't. So now bank runs tend to involve professional rather than retail counterparties.

### 4.1.5 How Financials Fail: Part 1

Financial institutions fail either because they cannot roll their funding – a liquidity crisis – or because they lose too much money – a solvency crisis. Sometimes rumours of losses precipitate a liquidity crisis. Here's why. If a firm loses a significant fraction of its capital, either because too many of the loans it has made default, or because the assets it has bought plunge in value, then the firm's credit quality falls. It is less safe because some of its loss absorption capability has been used. Further losses would push it closer to insolvency. This causes people to be less willing to assist in funding the bank. Its credit rating may be cut and the interest it has to pay to borrow money will increase. This in turn increases its costs and so reduces its profitability. The firm faces an uphill struggle to restore profitability and rebuild its capital. This brings the '6' part of the 3–6–3 rule into focus: if you are going to make and keep a loan, then make sure you earn your 6% from someone who is very likely to pay you back. Even the rumour that some of your borrowers won't pay you back could have highly adverse consequences for you.

### 4.1.6 The Uninformative Nature of NII

Old style banks have one enormous advantage compared with many other types of financial institution. This comes from their accounting. Consider a 3–6–3 bank making a ten year £1M loan at 6% interest. The borrower pays the bank £60,000 a year. The loan is funded with £100,000 of equity and £900,000 of deposits. In keeping with the 3–6–3 rule, the bank pays 3% interest on the deposits or £27,000. This leaves £33,000 of income. This income is called *net interest income* or NII because it comes from the difference between interest received on the loan and the interest paid on the funding.

The return on equity is $33/100$ or 33% (before tax). Where does this 33% ROE come from? There are two components:

- Income from taking *default risk*. The loan counterparty might not pay the bank back. The bank charges them extra interest to compensate for this risk.

- Income from *the mismatch in the term of funding*. To be completely term matched, the bank should have funded the loan by borrowing for ten years. It didn't, instead taking demand deposits. It has to pay less interest on those deposits than on a ten year loan because the Libor curve points up: ten year money is more expensive than overnight money.

Figure 4.2 shows these two components of NII: here the bank's cost of ten year money is 5.15%, so the term structure mismatch part is 2.15% of £900,000 or £19,350. The default component is the rest: £13,650.

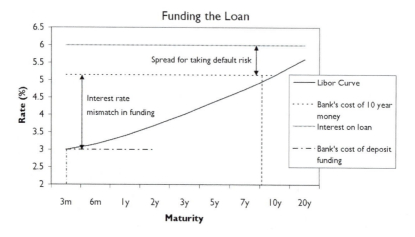

**Figure 4.2**: The components of net interest income

The lack of visibility of the components of NII is extremely helpful for banks. Their investors cannot see how much net interest income comes from which risk. It means that when it is hard to make much money from default risk – perhaps because money is cheap, as in the Boom years – banks can compensate to some extent by taking more funding liquidity risk. It is very difficult to understand the balance between these two risks as banks only have to disclose their total NII, not the components of it.

## 4.2   Liability Liquidity

Whenever one of its liabilities is due, a firm has to pay it. Very often raising the money requires another liability to be raised, if not precisely when the first one is due, then at least soon before. This process of rolling funding is vital to banks being able to continue lending, so it is worth understanding the main classes of liability and their liquidity characteristics.

### 4.2.1   Retail Deposits

Some people consider retail deposits to be the best source of debt funding. This is because retail depositors are not very confidence-sensitive: providing they have the benefit of effective deposit insurance, they tend not to withdraw their money when a bank gets into trouble. Thus even though contractually many retail deposits are demand money, in practice they tend to be rather stable[102]. This makes them good sources of funding for long term lending. Historically a large percentage of bank funding has come from this route.

### 4.2.2 Wholesale and Interbank Borrowing

If you cannot attract sufficient retail deposits to fund your assets, then you can try to plug the gap by taking deposits from other firms. Wholesale deposits are larger than retail deposits. However because they do not benefit from deposit insurance, they tend to be much more confidence-sensitive: providing there is no bad news, you can attract them; but if things start to go wrong, this source of funding can disappear very fast.

### 4.2.3 The Growth of Non-Deposit Funding

How can old-style banks make more money? There are basically three ways:

- Pay less for funding (i.e. turn 3–6–3 into 2–6–3);
- Originate higher spread assets (i.e. turn 3–6–3 into 3–7–3);
- Originate more assets.

The first is difficult. There is a lot of competition within the financial system for funding, and offering lower interest deposit accounts than your competitors is not a winning strategy. The second typically involves taking more risk: higher spread assets often pay more because they are more likely to default. Thus decreasing the quality of assets makes the institution riskier; and that may, in turn, increase its funding costs.

That just leaves increasing the size of the balance sheet: make more mortgages; buy more securities. The only two impediments to this are capital and funding. If the bank has enough of the former – or can figure out how to take risk without much more capital being required by its regulators – then all it has to do is fund the new assets.

Unfortunately there are not enough deposits to help here. Beyond a certain point, a country's banking system can only grow its assets further by borrowing: it has exhausted all the deposits that are available. In the UK, for instance, this happened around 2001. Before that point the major UK banks' customer lending was roughly equal to their total customer borrowing: in other words, across the system as a whole, deposits funded loans. After that lending grew faster than deposits. By the end of 2007 the gap was £300 billion, a reasonably substantial amount to borrow[103], and a sign of increasing funding liquidity risk in the banking system.

### 4.2.4   Commercial Paper

One of the most dangerous sources of borrowing is *commercial paper* or CP. This is because CP is a short term, institutional market. Debt is issued for less than a year – sometimes as little as a month – and in large size. Once it has been issued, CP tends not to be very liquid. Therefore most CP is issued by high quality companies. Any hint of difficulty tends to be very negative for a company's ability to issue CP: investors are very picky. Thus a firm can issue three month CP and roll it every quarter for some years, then suddenly find that CP investors don't want to buy. If this happens, the firm has to find an alternative source of funding quickly.

### 4.2.5   Term Debt

It is possible to borrow for the long term. Many firms issue bonds with terms from one to thirty years, occasionally even more. This *term debt* does not have to be refinanced for a long period, and hence it offers certainty of funding for some years. The disadvantage is that term debt is expensive.

### 4.2.6   Lines of Credit

Borrowing money and promising to pay it back is not the only way of managing liquidity needs. Aside from the obvious[104], another approach is to pay a fee for *the right* to borrow money if needed. This is a *line of credit*. It gives some flexibility, allowing a firm to take more funding liquidity risk knowing that if they get into trouble, they can draw down on the line.

The problem with lines is that drawing on them is a negative sign which can sap confidence. And that in turns means it will have more difficulty with funding in future. Thus for instance Countrywide did not draw on their line until their liquidity problems had become acute. This is a recurrent problem with contingent liquidity: a firm might have the right to use it, but by so doing, it can do more harm than good.

### 4.2.7   Repurchase Agreement

If you cannot borrow money just on the basis of your good name, then you have to give collateral. This is familiar from a mortgage. Similar kinds of borrowing happen on the capital markets. The most important form is the repurchase agreement or *repo*. Here a firm takes out a loan and pledges a security – usually a bond – as collateral. Repos are either overnight or term, with overnight being both cheapest and most common. These overnight repos are frequently rolling, that is both parties agree that each day the financing will be extended unless one of them decides to terminate.

A repo works roughly as follows.

- The party needing funds identifies a bond which can act as collateral. Typically this is a reasonably good quality security.

- The party identifies a potential funder for this bond. Terms are agreed including: the term of the repo; whether it will roll or not; the repo rate, i.e. the interest rate to be paid on the borrowing; and the value of the bond to be pledged against the loan.

- At the same time the *repo haircut* is discussed. The loan is smaller than the value of the bond by the amount of the haircut, so with a 5% haircut, if $100M of bonds were pledged, then only $95M would be lent. Thus 100% minus the haircut is the repo LTV.

- The party needing funding hands over the bond and gets the cash.

- At termination of the repo they buy back the bond, paying the loan amount plus the agreed interest.

This form of borrowing is particularly important to the broker/dealers. They cannot use deposits to fund their assets as they are not banks so do not have them. Furthermore they need to carry significant inventory of bonds so that they have a wide range available to sell to their clients just like any other type of retailer[105]. These bonds have to be funded. The easiest and cheapest place to do that is the repo market. Thus the large broker/dealers typically got tens of billions of funding via repo.

### 4.2.8  The Risk of Repo to the Lender

All the ideas needed to analyse the risk of a repo have already been discussed: after all, a mortgage is lending with collateral, just like a repo, so repo risk analysis should be just like mortgage risk analysis. Indeed, the two are directly comparable: just as in a mortgage one wants low LTV and high FICOs, so in a repo one wants a high haircut and a counterparty of good credit quality.

In both cases, too, there is a combination of risks one should consider: even if the counterparty is good quality now, could falling collateral value make them more likely to default? This happens in non-recourse mortgages

| | Mortgage | Repo |
|---|---|---|
| Quality of borrower | FICO | Credit quality |
| Collateral coverage | Better with low LTVs | Better with high haircuts |
| 'Wrong way' risk | Falling house prices may make default more likely | Falling bond prices may make default more likely |

Figure 4.3: Mortgage and repo risks compared

where a borrower with negative equity is more likely to default: in a repo it can happen if the borrower's credit quality is highly correlated with the value of the bond they are pledging as collateral[106]. This is sometimes known as *wrong way* risk.

The thing that distinguishes repo from mortgage lending is that different bonds can have very different price behaviours. The best short term government bonds tend to have rather stable prices as they are very good credit quality and changes in interest rates do not affect them that much; on the other hand, long dated bonds issued by lower credit quality corporates can jump substantially in value if the issuer's fortunes change.

For an overnight or rolling repo, we only have exposure to one day's worth of price change, so jumps in value are a concern. Therefore government bonds are usually considered safe collateral in a repo – and thus have a low haircut – whereas corporate bonds, particularly junk bonds, are much less good. Often repoing them will involve a high haircut and a higher cost of borrowing if it is available at all.

### 4.2.9    Collateralised Funding in the ODM

Repo is a form of secured borrowing where the collateral is a bond. ODM firms cannot use the repo market directly to fund their assets as they are not bonds: rather, they are things like mortgages or other kinds of loans. The solution is the issuance of ABS: in the ODM, institutions reduce their need for funding by securitising assets, that is by selling them to SPVs which in turn use the assets to back ABS. For example, as we discussed in section 3.1.7, mortgage banks used ABS issuance to free up funds allowing them to conduct further lending.

### 4.2.10    Managing Liability Liquidity

Some financial risks are easy to manage because they are quantifiable. You decide how much you can afford to lose and that determines how much risk you take. The process is not foolproof, but most of the time it appears straightforward. Funding liquidity risk is not like that as you cannot quantify confidence. Therefore firms' management of funding liquidity risk tends to be based on broad principles[107] which clearly mitigate the risks. The right balance between cost of funding and funding liquidity risk however remains a matter of judgement.

*Diversification* of funding is one important principle. Firms use a variety of different markets – CP, interbank, retail deposits, commercial deposits, and various forms of term debt – in the hope that a problem in one of them may not spread to the others. They also use a spread of maturities so that if there is a widespread issue at one point in time, the firm will not have too much debt to refinance during a problematic period. Unfortunately while that is a good concept, many firms thought that the period of difficulties would last for months rather than years. Now, as they face their second year of the Credit Crunch, some of them are finding that the cost of funds is still high and their ability to find sufficient funding is questionable.

*Limits on term structure mismatches* are also important. These constrain the amount of funding that has to be rolled over during any period and hence at least spread any problem out over time.

Firms also sometimes try to restrict the *total percentage of confidence-sensitive funding*. A large retail bank for instance might not wish to have more than 20% of its funding coming from non-deposit sources. Sadly however this is not possible for non-banks: a broker/dealer might have 75% or more of its funding coming from confidence-sensitive sources because it has no alternative.

Finally, firms need a plan for managing stresses. Some funding crises are market wide – as in the Crunch – others are localised. But in either case the firm should have a process for dealing with these events. The liquidity crisis plan recognises that once funding becomes too difficult the firm cannot just keep trying to raise money by borrowing.

### 4.2.11 How Financials Fail: Part 2

Earlier we saw how losses can lead to a failure of confidence. If a loss of confidence coincides with the need to borrow, then a firm may be unable to roll its liabilities. At this point attention switches to the asset side of the balance sheet. If you cannot borrow, one way to raise funds is to sell assets. But in order to be able to do that, the assets need to be liquid. In a crisis, asset liquidity decreases.

Thus there is the potential for a double whammy: decreased asset liquidity; and increased funding costs combined with decreased availability of funding.

It is well-known that this can happen. But in normal markets, this toxic combination of asset liquidity and funding liquidity risks is rare. During asset price booms there is plenty of money to be made for those financial institutions who are willing to provide credit. Three factors can increase a firm's profits:

- *More funding liquidity risk*, since that allows a bigger pick-up between borrowing short term and lending long;
- *More asset liquidity risk*, since illiquid assets earn a higher return than liquid ones;
- *More leverage* in the firm's capital structure, since equity funding is more expensive than debt.

There is another pressure towards more leverage too: equity holders want a high leverage structure as that means that excess profits are distributed between a small number of shares, and hence the earnings per share are high. In contrast, holders of the firm's debt want a low leverage structure so that the firm can absorb lots of losses and ride out periods of illiquidity without a loss of confidence. But the firm is owned by and operated for equity holders. All of these pressures, then, lead towards more risk.

There are two defences against this. One is that higher leverage firms have lower quality debt, and thus if a firm's leverage increases too much, its credit spread will increase and it becomes more expensive to borrow. But this tends not to be an absolute effect: rather it is if a firm's leverage becomes too high *relative to its peers* its credit spread goes up. If the average leverage of all firms increases, average credit spreads do not necessarily change.

The other protection is supposed to be regulation. Regulators try – not always successfully – to moderate firms' drift towards increasing risk. When they fail, as they did in the Boom years, firms become riskier as their effective leverage rises.

When leverage is too high, losses comparable with capital are possible. If this happens (or even if it is thought likely to happen) and there is too much funding liquidity risk, the firm then becomes a forced seller. One firm in this situation is a problem for that firm, but not for the market. The presence of many forced sellers at once, though, impacts liquidity. Then there are not enough buyers to absorb the supply of assets on the market. Thus a cycle of selling – losses – illiquidity – selling develops.

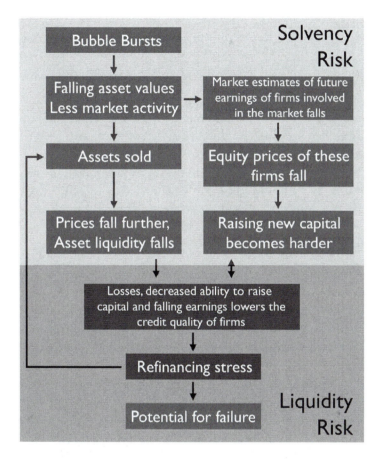

**Figure 4.4:** The vicious circle of solvency risk and liquidity risk for lever-aged holders of assets as prices fall

This gives rise to a vicious circle as illustrated in figure 4.4. If this cycle gets going, plunging confidence, difficulty of raising funds, and falling asset liquidity makes it very hard for firms to help themselves. The Central Bank may need to intervene.

## 4.3   Central Banks

The Federal Reserve System has a central rôle in the U.S. economy. Similarly the European Central Bank (ECB) is key for the Euro-zone countries, and the Bank of England rules Sterling. Their operations have a dramatic effect on the financial markets, for good or ill.

All of the major Central Banks have been very active in trying to alleviate problems caused by the Crunch. But what do they do, exactly, and why?

### 4.3.1 What's Good?

The Central Banks share various versions of the same target: sustainable economic growth with price stability. Thus for instance the Federal Reserve Act specifies that the FED should seek "to promote effectively the goals of maximum employment, stable prices, and moderate long-term interest rates"[108]. For some Central Banks there is an explicit inflation target – the Bank of England for instance is currently mandated to keep inflation between 1% and 3%[109] – while for others the control of inflation is implicit. But in all cases Central Banks are supposed to keep long term inflation down. They do this using *monetary policy*.

Stable economic growth requires a stable financial system. If banks do not lend, then businesses cannot borrow to invest. Confidence in the financial system is key to keeping funding costs low and hence allowing banks to lend at low absolute interest rates. Therefore Central Banks also take an interest in *financial stability*. Their specific rôles here vary: the FED, for instance, is responsible for regulating some (but not all) financial institutions; whereas the Bank of England merely has "responsibility for the stability of the financial system as a whole" without actually regulating anyone[110].

### 4.3.2 Financial Stability

There are a number of different aspects to ideas of financial stability:

- *The safety of deposits.* Depositors may not suffer in a bank failure due to deposit insurance. Governments however do, because they (or their deposit insurance corporations) have to make good the shortfall. Therefore financial stability aims to minimise the number of failures of large banks.

- *The availability of credit.* The economy needs credit to grow, so financial stability requires that the financial system provides a ready and equitable supply of credit at a fair price.

- *Orderly financial markets.* Wild price swings are not conducive to financial stability. Therefore some commentators, although by no means all, consider that Central Banks have a rôle to play in controlling asset price volatility[111].

- *Failures are localised.* Financial institutions, particularly small financial institutions, fail all the time[112]. It is unrealistic to try to prevent them. However it is reasonable to ask that if a firm fails, it does not endanger other institutions when it goes down.

The term *systemic risk* is often used to describe the danger that a failure of one institution may cause problems for others. Narrowly, systemic risk is simply direct credit risk: one institution may have borrowed from another, and if it fails then the lender will suffer a loss. However there are other vectors of systems risk including asset prices – forced liquidation of a bankrupt firm's assets may cause losses in fair value for other firms – and losses of confidence in the system as a whole affecting funding.

### 4.3.3 Monetary Policy

Monetary policy is the process by which Central Banks try to meet their economic growth and stability objectives. They control money. In the distant past, money meant physical currency. But these days most money is not in that form: rather it is in deposits, securities, and other liquid assets.

The Central Bank influences demand in the economy in two ways:

- Controlling the supply of money; and/or
- Adjusting the demand for money by altering interest rates.

Changes in monetary policy are usually explained in terms of the control of interest rates, specifically a short term interest rate known as the *short rate*. In its crudest version, the idea is that if inflation starts to rise too far, interest rates are raised to choke off demand: if growth is too sluggish, rates are lowered to stimulate demand. Thus lowering rates stimulates the economy. But how does the Central Bank actually control the short rate?

### 4.3.4 The Conduct of Monetary Policy

The major monetary policy tool used by the major Central Banks is a form of repo[113]. The use of these repos is known as *open market operations*. They allow the Central Bank to control the supply and cost of credit to banks as follows:

- The Central Bank – or an affiliated committee – decides on the desired short rate based on economic conditions;
- If the banks want cash, they can get it from the Central Bank. The Central Bank is a monopoly supplier of new money. It does this by allowing banks to approach it for loans at a suitable rate. This loan is in the form of a repo: the banks pledge bonds to the Central Bank and get (freshly created) cash in exchange. This process is sometimes known as *going to the window*[114].

In a well-functioning financial system the banks then pass this interest rate on to the rest of the economy: if their funding costs go down, because the Central Bank has lowered the rate at the window, then the banks can charge their customers less.

The Central Bank therefore controls the short rate by controlling the supply of money at the window.

### 4.3.5   The Window and Collateral at the Window

Various Central Banks have various types of open market operation. We will examine the details of one, the ECB's *main refinancing operation*. Every week the ECB offers one week repos to eligible banks. To be able to go the ECB's window, a firm must be a regulated, financially sound Euro area bank. It must also hold minimum reserves at the ECB: these are deposits that the ECB requires banks to leave with it: the larger the bank, the bigger the reserve requirements[115].

The actual refinancing operation is carried out by the national banks of the Eurozone system. Thus for instance an eligible French bank would deal with the Bank of France. First the ECB announces the operation. Thus for instance on 8th September 2008 the ECB announced a regular main refinancing operation. Banks were invited to submit bids to repo bonds. The minimum rate allowed for bids was 4.25%. This was the ECB's target for the short rate in Euros. Counterparties had until 9.30am on the 9th to submit bids. These are in minimum amounts of a million Euros, and contain the offer of an interest rate. Thus a French bank might submit a bid to the Bank of France where it offers to enter into a repo transaction, receiving ten million Euros and pledging bonds worth ten million Euros plus the haircut. The bank will undertake to buy the bonds back in a week for a higher price, corresponding to an (annualised) interest rate of 4.3%.

(A similar decentralisation takes place in the American system: the Board of Governors of the Federal Reserve system oversees policy, but the actual implementation takes place via the twelve district Federal Reserve Banks. A large share of the burden falls to the New York FED.)

Banks cannot bid lower than the minimum, but they can bid higher if they are desperate for cash: in fact banks can submit up to ten bids at different rates, giving them more control over how much cash they get and how much they pay for it. The ECB lists all the bids in order of the interest rates offered and fills the highest rates first. Typically a fixed amount of liquidity is available – although the ECB has some discretion here – so banks which only offer the minimum rate run the risk of not receiving as much cash as they want.

The ECB's operation settles the next day, so bids received on the 9th September give the banks cash on the 10th and this has to be repaid on the 17th. For this particular operation the ECB lent the Eurozone banking system €176.5B at a weighted average rate of 4.41%.

### 4.3.6  Emergency Funding: The Lender of Last Resort

Bank collapses upset people. Even if they get all their money back – thanks to effective deposit protection – they do not like their bank to fail. And if they don't get their money back, people are very very upset. Therefore for both political and economic reasons governments go to considerable lengths to avoid outright bank failures.

Systemic risk is not confined to banks alone: Bear Stearns was not a bank, and yet it was judged to pose enough systemic risk that the FED intervened to prevent its failure. Central Banks do not often allow systemically important firms to go bankrupt: instead the firm is taken over and emergency funds are injected by the Central Bank as *the lender of last resort*.

Access to the Central Bank as lender of last resort is part of the unique bargain of banking. Banks have advantages which no other class of firms have: they take depositors' money and lend it on freely; they enjoy deposit protection which makes it easy for them to attract funding; they are highly leveraged, to the benefit of their shareholders. But the price of these freedoms is twofold: regulation, including (some) limits on leverage; and rules which permit the state to intervene before the bank has failed, protecting depositors but depriving shareholders of control. Whereas an ordinary corporation can slide close to disaster and perhaps recover, a bank is not allowed very close to the edge before it is nationalised. This is part of the nature of banks and should be understood by anyone who buys shares in them.

How regulators choose whether to save a firm is a delicate and controversial business. No Central Bank wants an institution to be sure that it will be rescued, as that may encourage more reckless risk taking. This is the *moral hazard* of Central Banking[116]. When problems do actually arise, however, the Central Bank must make a decision on the consequences of allowing a bankruptcy versus the cost to the taxpayer of rescuing the firm and the moral hazard involved. Typically small non-banks which pose no systemic risk will be allowed to go bankrupt, whereas banks will be taken over as part of the deposit protection regime. Shareholders in these banks typically lose everything, and senior creditors may well receive only whatever is left after depositors have been made whole.

Large firms are sometimes not allowed to get to this point. The systemic risk even of an orderly supervised wind-down is viewed as too large. These *too big to fail* banks are taken over and run as going concerns. Arguably there is significant moral hazard in this process. However national champion banks are often (but not always) supervised more rigorously than smaller institutions: whether that makes up for an implicit guarantee of state support is a political question[117].

A problem comes in deciding what to do about large non-banks. Central bankers tend to be pragmatic here. For instance when the large hedge fund LTCM failed in 1998, the FED arranged a bail-out. Admittedly it did not lend to LTCM, but it ensured that the fund was wound down in an orderly fashion[118]. In the latter stages of the Crunch, Central Banks certainly demonstrated a good deal of flexibility.

It is worth noting that while both equity and debt holders in a firm take risk, equity holders take the bigger risks in expectation of bigger rewards, and so rescuing them involves more moral hazard. Saving a bank often means that the senior debt holders are paid in full. There is some moral hazard here, but not nearly as much as when equity holders receive a bail-out.

## 4.4    Central Bank Policy in a Crunch

Central banking orthodoxy in the 1990s was that whenever the economy got into trouble, you cut rates. The thinking was that if you judged things correctly, you could provide a sufficient stimulus to keep the economy growing without risking inflation. But if you cut too far or for too long – as some economists think Alan Greenspan did[119] – then you risk an asset price bubble.

When the severity of the Crunch first became apparent, Central Banks reacted in the usual way: they cut rates. It didn't work. To see why – and what the Central Banks did next – we need to look at all the policy levers available in monetary policy.

### 4.4.1    Engineering the Repo

There are four things a Central Bank can control about its window:

- How much financing it is willing to provide at a given time;
- What collateral it is willing to take;
- Who they are willing to lend to; and finally
- The rate they are willing to lend at.

The first three of these are just as important as the rate.

### 4.4.2 The Quantity Theory of Stability

In normal conditions, relatively small changes in the amount of money the Central Bank makes available can affect conditions. There is enough liquidity for the financial system to function, and a little more makes it go faster. But when there is a financial crisis it takes hundreds of billions to make a difference.

This is because confidence disappears in a crisis. Firms cannot borrow as easily either on an unsecured basis or using collateral. In particular the haircuts for risky assets rise, making it more expensive to finance assets. There is a liquidity drought as those firms with liquidity keep it in case they need it, and those without scramble to finance themselves anywhere they can. In this situation a few tens of billions more from the Central Bank does not make much difference. To keep banks lending – and hence to prevent credit rationing from affecting the broad economy – the Central Bank has to be willing to provide enough cash to allow banks to deleverage themselves sufficiently to restore confidence.

Notice that liquidity directly affects repo funding. Repos are only a safe form of lending if the lender can be sure that the collateral can be sold for more than the value of the loan. If the collateral is illiquid, then it is hard to value. This means that it is hard to be confident that the loan is fully collateralised. Interbank repo funding therefore tends to dry up when bond liquidity falls.

### 4.4.3 The Pawnbroker of Last Resort

In a crisis, risky assets become hard to finance. Thus for instance during the Crunch, ABS not only fell in price: repo haircuts rose and the term of available repos fell. Players with access to lavish funding could take their own view on whether prices had fallen too far. But those without that advantage – including highly leveraged non-banks such as hedge funds – had to sell. If you cannot fund a security, you cannot keep it.

The increased funding costs for risky assets caused the non-default components of the credit spread to increase. But these can be directly manipulated by Central Banks. If ABS are acceptable collateral at the window, banks can finance these assets cheaply. That means they are not incentivised to sell them into a falling market. It also allows Banks to use their deposits to make new loans, as these funds are not diverted into funding old assets.

This medicine needs to be used rather cautiously. For a bank, Central Bank funding is a little like heroin for a drug addict. Give them too much of it, and they become reliant on it. Without the discipline of raising most of their funds in the market, banks are receiving a state subsidy. And they have little incentive to control the risk of the assets they acquire if they know they will always be able to finance them with the Central Bank. The availability of Central Bank funding for risky assets helps to stabilise the financial system in a crisis, but it should not be available forever.

### 4.4.4    Throwing Open the Clubhouse Doors

Who should the Central Bank allow to approach its window? The old-fashioned answer was 'only large, well-capitalised, local banks'. The Central Bank has a duty to the banking system, and in this view the banking system is the banks. Moreover minimising the value of claims on the government's deposit protection scheme involves protecting deposit takers, and those are the banks. Why should the Central Bank deal with anyone else?

The answer is financial stability. There are many non-banks that have a prominent place in the financial system including broker/dealers, some insurance companies, and hedge funds. The biggest financial markets crisis of the 1990s involved a hedge fund, LTCM. Many of the prominent companies that suffered distress in the Crunch – including Bear Stearns, Freddie & Fannie, Lehman Brothers, Merrill Lynch, and Countrywide Financial – were not banks. Moreover the banks have significant relationships with non-banks and significant dealings in markets where non-banks intermediate.

Financial stability considerations may therefore lead the Central Bank to allowing non-banks access to its liquidity.

## 4.5    A Tale of Two Central Banks

The two most important Central Banks for the financial system are the FED and the ECB simply because U.S. dollars and Euros dominate international banking, and the U.S. and Eurozone between them are home to many of the most important institutions. What did the FED and the ECB do during the Crunch?

### 4.5.1    Before the Stress

In early 2007, the two leading Central Banks had two rather different approaches[120]:

- The FED conducted monetary policy both through very short term lending to depositary institutions – a process known as use of the *discount window* – and through repo. The permitted counterparties for the latter are known as *primary dealers*: they include both some large broker/dealers and the securities dealing subsidiaries of the largest banks. While repo with the primary dealers was the most important tool in monetary policy, it was somewhat indirect: the FED controlled the supply and cost of funding to the primary dealers, and these firms distributed it on to the rest of the financial system via the repo market. The discount window acted as a safety valve, but this meant that there was a stigma associated with borrowing at it. Banks were reluctant to use it as they did not want to provide any confidence-shaking signal to the market.

- The FED's indirect mechanism – using the primary dealers to on-lend to the banks – reflects the uniquely American financial system where banks were not historically the largest players in the securities markets. This is not the case in the Eurozone, so the ECB's main refinancing operations are carried out directly with banks. Another difference is that the ECB accepted a wider range of collateral than the FED, although its haircuts for lower quality assets were significant[121].

### 4.5.2 Rate Cuts Fail: the TAF is Introduced

The FED's first attempts to ensure financial stability in the early days of the crisis involved cutting rates. Thus the FED funds rate was cut on the 18th September 2007, and again on 31st October and the 11th December. These cuts alone were not enough: banks' real cost of funds remained high as conditions became more and more difficult in the interbank and collateralised financing markets.

The FED's response was to turn to the other levers of policy. It began by instituting a new facility, the *Term Auction Facility* or TAF, on 12th December 2007. Here depositary institutions could bid for funds directly with the FED without having to use the discount window. Funding is allocated by auction, much like the ECB main refinancing operation. This allowed the FED to use modern repo-style operations with a wider range of counterparties than just the primary dealers. Terms were also somewhat longer than the usual very short term Central Bank lending: one month and, latterly, three month TAF auctions were instituted[122].

The sums provided via the TAF were reasonably significant: by September 2008, $75B of one month credit and $25B of three month credit were being offered at each auction, to a total outstanding of $150B, and by early 2009, $150B was being offered at a single auction.

The TAF was a qualified success. The Libor/FED funds spread reduced, and banks obtained liquidity at a rate lower than that at the discount window and without the associated stigma.

### 4.5.3   Financing Not Liquidity: the TSLF and the PDCF

The TAF helped to address the provision of liquidity: it gave firms some access to cash. But as the Crunch developed, the interbank market started to dry up again. The TAF was not providing enough funds, nor providing them widely enough to keep the financial system functioning well. Liquidity premiums were the culprit here: market participants had become extremely reluctant to provide funding to each other for risky assets. They were hoarding the cash provided by the TAF rather than on-lending it. Some parties were getting heavily squeezed, particularly those not able to go to the FED – because they were not primary dealers – or who wanted to get funding for assets that were too low quality to be eligible under the TAF.

These difficulties show that the *quantity* of funding alone is not the only determinant of market distress. Turning the cash spigot on will not necessarily make things better if the market does not distribute this cash to all those that need funding. The FED therefore instituted two new facilities aimed at reducing stress in the collateralised funding markets and hence helping those who were reliant on repo funding for risky assets. These two facilities were the *Term Securities Lending Facility*, instituted on the 11th March 2008, and the *Primary Dealer Credit Facility*, brought in a few days later[123].

As the name indicates, the TSLF is a term facility: it offers one month funding. The facility is a *two way repo*: the counterparty gives the FED a lower quality asset, and the FED gives the counterparty a treasury bond. This treasury bond can then be financed in the usual way, as the repo market remains open for high quality collateral. Thus firms with lower quality hard-to-finance bonds swap them for assets which can be easily financed. The TSLF therefore does not change the money supply, it simply liquefies an illiquid bond, and thus reduces the non-default-related component of the credit spread of eligible bonds.

The PDCF is simpler: essentially, it extended access to the discount window to the primary dealers on more or less equal terms with depositary institutions.

Thus the FED found it necessary to widen access to its liquidity beyond member banks. This reflects the failure of the banks to distribute cash around the financial system in the Crunch.

### 4.5.4 The Largest SIVs in the World

As the Crunch progressed, the collateral eligible for use in the TSLF widened. By late 2008, the FED was willing to finance any investment grade bond, including ABS. The TSLF therefore became complementary to the TAF: the former was about guaranteeing funding for assets; the latter was about distributing cash.

The ECB did not need a new facility like the TSLF: even before the Crunch, they already accepted a wide range of collateral at the window. Indeed, so generous was the ECB's collateral list that some banks started to create assets specifically for use with the ECB. This left a Central Bank with a dilemma: clearly offering Euros for bonds that had never been sold, and hence there was no reason to believe in the credit quality of, was imprudent. Equally a crackdown on the range of eligible collateral could cause a liquidity crisis in the Eurozone[124].

### 4.5.5 The Evolution of Central Bank Policy

The old model of monetary policy focussed on the money supply. Here Central Banks were seen as controlling the amount of cash banks received. They took as little risk as possible, and specifically did not involve themselves in the asset markets. The Crunch comprehensively changed that model. It is now clear that collateralised funding markets are as important for financial stability as the supply of funds to depositary institutions: perhaps more so. Central Banks cannot simply provide short term funding to banks and hope for the best. In times of stress, they need to offer both cash *and* asset financing. Reluctantly they have taken more and more risk: on the 14th September 2008, for instance, the FED extended the TSLF to all investment grade bonds and increased its size to $200B.

The risks to the Central Bank of this new policy are slight compared with its rewards. If a counterparty fails, then it may lose the difference between the liquidation value of the collateral and the loan amount. But even the three month TSLF is fairly short term, and the haircuts are considerable. The risk seems worth taking given the benefit to financial stability. In times of market stress, guaranteed funding for risky bonds is vital in calming the market. This new approach reflects the importance of the bond markets and asset financing to the banking system. It is known as *monetary policy on the asset side of the balance sheet*, since the Central Bank is acquiring assets rather than changing the structure and cost of its liabilities as in traditional discount window lending[125].

The problem Central Bankers face is balancing financial stability with moral hazard. In good times there is no justification for allowing certain favoured clients of the Central Bank – depositary institutions – to finance assets cheaply while others – hedge funds – have to fend for themselves. The mitigation of systemic risk probably justifies innovations like the TAF and the TSLF, but these cannot last forever. At some point the Central Banks will have to begin the detox. Indeed, faced with increasing amounts of dubious collateral, the ECB has already begun this process. On 4th September 2008 it announced a reduction in the range of eligible collateral and an increase in some haircuts from February 2009[126]. The FED, with a bigger problem to address, has done nothing yet.

If they go too far too fast, the Central Banks risk turmoil in the money markets. Monetary policy on the asset side of the Central Banks' balance sheet may be reduced once the Crunch is over, but it will have a place in the Central Banker's toolkit in addressing future market crises.

## 4.6    A Twenty First Century Run: Northern Rock

The scenes on some British High Streets early on the morning of 14th September 2007 were extraordinary. By 7am there were long queues outside many of the branches of Northern Rock. The first run on a British bank since 1866 was underway[127]. The U.S., despite a number of bank failures, had not produced scenes like these: what went wrong?

### 4.6.1    *What Was Northern Rock?*

Northern Rock was a bank based in the North East of England[128]. It was formerly a building society, so it naturally had a focus on mortgage lending. Like Countrywide, it had used a market in which property prices were growing fast to build its loan book, and by the end of 2006 it had over £100 billion of assets, nearly 90% of which were residential mortgages.

Unlike Countrywide, Northern Rock did at least have significant funding from deposits. But the bank's deposits alone were not sufficient to fund all the mortgages Northern Rock could make, and so it started to distribute some of its loans, selling packages of mortgage risk through a SPV, Granite. By the start of 2007, ABS issued by Granite provided roughly 50% of Northern Rock's funding[129], with retail deposits providing only around 25%.

The first warnings of the Crunch in late 2006 and early 2007 were ignored by the Bank. The Northern Rock mortgage book was expanded by £10 billion in the first half of 2007. The bank was active not just in good quality retail lending, but also in buy-to-let mortgages and high LTV loans[130].

### 4.6.2   Funding Crunch

Northern Rock's funding model gave it some problems. As the Crunch developed, buyers of mortgage-backed assets disappeared. Northern Rock did not have a particularly high credit rating: it was rated single A+. Therefore both its cost of issuing unsecured term debt and its ability to do it in multi-billion dollar amounts were a problem. Without the ability to refinance its maturing debt, the firm would fail.

Northern Rock tried to raise funds in the wholesale market. However by this point contagion had overtaken the debt markets: the problems in the mortgage market had caused widespread distrust of banks, and Northern Rock's sudden appearance in the wholesale markets asking for billions of pounds of funding was not successful. Northern Rock did not have significant funding available via lines of credit, unlike Countrywide, so it was forced to apply to the lender of last resort. The first discussions with the Bank of England took place on 14th August, and by 13th September Northern Rock was receiving emergency funding.

### 4.6.3   How Risky Was Northern Rock's Funding Strategy?

Northern Rock failed due to its liquidity risk. Therefore its funding strategy, in retrospect, was too risky. But was that obvious before the fact? Some people think so: one expert called them 'reckless'[131]. The reality without 20-20 hindsight is less clear.

One issue is that since the ODM became widespread there had never been a buyer's strike in the ABS markets. There was no reason to believe that this could happen. Moreover, even if Northern Rock was unable to sell its ABS, it had a back-up strategy: going to the wholesale funding markets. The fact that both these markets closed to the bank more or less at once was difficult to foresee. Even then Northern Rock had a further piece of bad luck in that it was just about to issue ABS from Granite when the markets closed. If contagion from the U.S. sub-prime market to the wider debt markets had taken a month longer, Northern Rock might well have survived. Moreover, while the percentage of its funding coming from securitisation was unusual, the fact of it was not. Many UK banks used a funding mix similar to Northern Rock's, but with a higher percentage of deposits. The firm did have a risky strategy, but not, I would suggest, one that was obviously deeply imprudent based on what was known in 2006.

### 4.6.4   Why Did The Man In The Street Run?

The queues in the street were caused by the unique and flawed design of the UK deposit protection regime. In the U.S., the FDIC guarantees retail bank deposits up to $100,000. Moreover payout is timely: when a bank fails, customers can typically access their deposits the next day. In the UK in 2007, the Bank of England only guaranteed 100% of the first £2,000 of deposits. 90% of the next £33,000 was also guaranteed. But if you have £35,000 of savings, losing £3,000 thanks to a partial guarantee is not a risk worth taking. Any depositors with more than £2,000 of total savings with Northern Rock rationally should have withdrawn their funds when the Bank's difficulties became clear. Moreover the Bank of England regime is not as timely as the FDIC's, so any depositors who were concerned about when they would get their money back were also rational in queuing.

The UK deposit protection regime, then, had at least two fatal flaws: it was not a complete guarantee, and it did not promise depositors quick access to their money. Subsequently some of this was recognised, and the UK deposit protection scheme now offers a 100% guarantee on amounts up to £50,000. The UK government still thinks however that making funds available one week after a bank failure 'would be an extremely challenging target in almost all circumstances'[132].

### 4.6.5   Uncertainty Persists

Another difference between the U.S. failures and the Northern Rock concerns the speed of intervention. One journalist has joked that Hank Paulson, U.S. Treasury secretary, should start taking weekends off, because whenever he works on a Sunday, a large financial services firm fails. But typically it does only take a weekend: on Friday, the firm is a going concern: by Sunday evening, the problem has been resolved.

The case of Northern Rock was one of long-drawn-out anguish. The UK regulatory authorities sought private sector bids for the bank during September and October 2007, hoping that a larger firm would have access to enough funds to be able to run Northern Rock's book profitably. The search continued through November. And December. And January. The final deadline for bids was the 4th February 2008. No acceptable offers were forthcoming, so Northern Rock was nationalised on the 17th February and run by a government-appointed team. An independent valuer has been appointed to decide what compensation shareholders should receive. Given that Northern Rock is worth nothing without government support, it is rather unclear why that number should be more than zero.

The structure of the UK regulatory regime is partly to blame for the slow motion nature of the Northern Rock failure. Political authority rests with the Treasury: the *Financial Services Authority* or FSA is the supervisor; and the Bank of England is the lender of last resort. Typically in any situation with two or more government bodies, there may be issues of cooperation and information exchange[133]. In this *tripartite system*, the FSA, as supervisor, had detailed information on Northern Rock but no money. The Bank of England could act as lender of last resort but did not want to unless there was no alternative, preferring a private sector bailout if one could be arranged. And the Treasury was in charge but had neither all the facts nor any means of acting. In those circumstances, a six month gap between the first problems and the final resolution is perhaps unprecedentedly speedy. To be fair, there have also been suggestions that European Union regulations may have prevented the secret negotiation of a speedy private sector solution. But certainly the comparison between Mr. Paulson's weekend bailouts and the UK's slow drift towards nationalisation is not to the latter's credit.

---

# Notes

[100]We are taking huge liberties here with two complicated things: capital, and accounting. For more details on the financial statements of financial services firms, see *Accounting Theory and Practice* by Glautier and Underdown, 2007. Both topics are discussed further in chapter 10.

[101]For more on liquidity risk see Chapters 8 and 9 of my book *Understanding Risk: The Theory and Practice of Financial Risk Management*.

[102]This is known as the *behavioural maturity* of deposits: it is usually much longer than their contractual maturity, and banks spend considerable effort in modelling this phenomenon. A longer discussion of this and related issues in liability management is given in *Banking System Liquidity: Developments and Issues* by Graeme Chaplin, Alison Emblow and Ian Michael, Bank of England Financial Stability Review December 2000 available at www.bankofengland.co.uk/publications/fsr/2000/fsr09art2.pdf. See also *Does Deposit Insurance Increase Banking System Stability?* by Ash Demirgüç-Kunt and Enrica Detragiache, World Bank 2000 available at www1.worldbank.org/finance/assets/images/depins08.pdf for a note of caution about the rôle of deposit protection in stabilising bank funding.

[103]For details see the *Bank of England Financial Stability Review*, April 2008, specifically the discussion around table 2.10. This document is available at www.bankofengland.co.uk/publications/fsr/2008/index.htm.

[104] Borrowing money and not promising to pay it back tends to go down badly in the bond markets.

[105] In fact the broker/dealers also carry significant inventory as pure proprietary positions or as a blend of client inventory and proprietary position.

[106] So in particular taking a Russian bond as collateral in a repo with a hedge fund 90% of whose assets are invested in Russia is not a great trade.

[107] See *Principles for Sound Liquidity Risk Management and Supervision*, Basel Committee on Banking Supervision 2008, available at www.bis.org/publ/bcbs138.htm for more details.

[108] A good introduction is given in *The Federal Reserve System: Purposes and Functions*, 2005 available at www.federalreserve.gov/pf/pf.htm.

[109] The current Bank of England inflation mandate is given in *The Remit of the Monetary Policy Committee*, Letter from Gordon Brown to Mervyn King, March 2007 available at www.bankofengland.co.uk/monetarypolicy/pdf/chancellorletter070321.pdf.

[110] The inherent contradiction in this is one of the reasons for the Northern Rock débâcle, as we shall see.

[111] For the positive case see *Asset Prices and Central Bank Policy* by Stephen Cecchetti and Hans Genberg, International Center for Monetary and Banking Studies. *Asset price bubbles and monetary policy* by J-C. Trichet, MAS lecture June 2005, available at www.ecb.eu/press/key/date/2005/html/sp050608.en.html contains an opposing argument from a Central Banker.

[112] The FDIC's Bank and Thrift Failure Report, for instance, details 29 U.S. failures between 2000 and 2005. See www2.fdic.gov/hsob/SelectRpt.asp?EntryTyp=30.

[113] For a more comprehensive account of open market operations, see The Federal Reserve System: Purposes and Functions previously cited, or the ECB's *The Implementation of Monetary Policy in the Euro Area*, September 2006, available at www.ecb.int/pub/pdf/other/gendoc2006en.pdf.

[114] Different Central Banks have different ways of conducting open market operations, and many of them have more than one. For instance, see *Forms of Federal Reserve Lending to Financial Institutions* available at www.ny.frb.org/markets/Forms_of_Fed_Lending.pdf for a summary of the various FED programmes. We will use the term 'going to the window' for any means of accessing cash from a Central Bank short of armed robbery.

[115] These reserve requirements were originally to ensure that banks had enough cash to meet redemptions of demand deposits. Nowadays some Central Banks – such as the Bank of England – have dispensed with reserve requirements entirely, while others use them as a low minimum liquidity standard with higher standards of liquidity risk management being imposed via other regulations.

[116] Moral hazard arises whenever a voluntary action by government bodies has the effect of lessening losses for risk takers.

[117] Certain investors, believing that the largest banks from Western Europe will never be allowed to fail, are active sellers of credit protection on them.

[118] Opinion differs as to whether the FED was justified in rescuing LTCM or not. For a broadly pro position, see *Some Lessons on the Rescue of Long-Term Capital Management* by Joseph Haubrich, Federal Reserve Bank of Cleveland Discussion Paper 2007 (available at www.clevelandfed.org/research/PolicyDis/pdp19.pdf), whereas *Too Big to Fail?: Long-Term Capital Management and the Federal Reserve* by Kevin Dowd, Cato Institute No. 52 1999 (available at www.cato.org/pubs/briefs/bp-052es.html) gives the opposing viewpoint.

[119] A balance of views is given in many recent articles. See for instance *Greenspan Stands His Ground* by Steven Mufson in the Washington Post, 21st March 2008. Mufson quotes criticism from Lee Hoskins, former FED chairman: "the Fed 'made money very cheap, and we began to see the whole leveraging process we see today' ". Jeffrey Sachs, Professor of Economics at Columbia, has a fuller account of the case against Greenspan in *The Roots of America's Financial Crisis*, Project Syndicate 2008, available at www.project-syndicate.org/commentary/ sachs139.

[120] For a broader comparison see *Monetary policy frameworks and central bank market operations*, Bank for International Settlements, December 2007, available at www.bis.org/publ/ mktc01.pdf.

[121] Compare Tables 4 and 7 of the ECB's The Implementation of Monetary Policy in the Euro Area with the FED's *Discount and PSR Collateral Margins Table* available at www.frbdiscount window.org/discountmargins.pdf.

[122] More details on the TAF can be found in the frequently asked questions lists at www. federalreserve.gov/monetarypolicy/taffaq.htm.

[123] More details on the TSLF and PDCF can be found in the frequently asked questions lists at www.newyorkfed.org/markets/tslf_faq.html and www.newyorkfed.org/markets/ pdcf_faq.html respectively.

[124] The Spanish banks in particular seemed to be relying on ECB funds to finance residential mortgages. Given the declining state of the Spanish property market, the ECB may have been concerned that withdrawing liquidity too fast could have caused a Spanish banking crisis.

[125] The coinage appears to be James Hamilton's: see *Monetary policy using the asset side of the Fed's balance sheet* available at www.econbrowser.com/archives/2007/12/monetary_ policy.html.

[126] See *Biennial review of the risk control measures in Eurosystem credit operations* available at www.ecb.int/press/pr/date/2008/html/pr080904_2.en.html. These decisions were subsequently revisited later in 2008.

[127] See *Banking Crisis Solutions Old and New* by Alistair Milne and Geoffrey Wood, Federal Reserve Bank of St. Louis Review, September 2008, available at research.stlouisfed.org/ publications/review/08/09/Milne.pdf.

[128] For more context and discussion, see the House of Commons Treasury Committee *The run on the Rock*, Fifth Report of Session 2007-08. This is available at www.publications. parliament.uk/pa/cm200708/cmselect/cmtreasy/56/5602.htm.

[129] A further 10% of Northern Rock's funding came from issuing bonds from the bank's balance sheet directly backed by a segregated pool of mortgages – *covered bonds*. Thus more than half the firm's funding was dependent on being able to sell mortgage-backed risk in some form or other.

[130] For an informal account of Northern Rock's growth and failure from the inside, see *The Fall of Northern Rock: An insider's story of Britain's biggest banking disaster* by Brian Walters, 2007.

[131] See *The Lessons from Northern Rock*, Willem Buiter, Financial Times 13th November 2007, available at blogs.ft.com/wolfforum/2007/11/the-lessons-fro.html/.

[132] See *The run on the Rock: Government Response to the Committee's Fifth Report of Session 2007-08*, Eleventh Special Report of Session 2007-08, available at www.publications. parliament.uk/pa/cm200708/cmselect/cmtreasy/918/918.pdf.

[133] Certainly relations between FSA and the Bank of England may not be optimal given that the Bank of England used to be the UK bank supervisor until 2000: FSA was given the power instead in part because the Bank did a less than universally applauded job of managing the failure of BCCI in 1991.

# Chapter 5

---

## The Crash of 1929 and its Legacy

*The money changers have fled from their high seats in the temple of our civilization. We may now restore that temple to the ancient truths... This Nation asks for action, and action now.*

Franklin D. Roosevelt

---

## Introduction

The last global financial disruption of a comparable magnitude to the Credit Crunch was the market crash of 1929 and the subsequent Great Depression. The similarities between the Crunch and these events are considerable, so it is appropriate to look back eighty years. Insights into the Crunch and how to tackle it abound in the events of 1929-1935.

There is also another reason to look back too: many of the institutions of American financial services are a result of legislation enacted during and after the Great Depression. In particular Fannie Mae – an important and troubled player in the mortgage market – dates from that time. To understand the U.S. regulatory framework, and to explain the nature of the American housing market, we need to go back to the innovations of Franklin Roosevelt's administration.

## 5.1  The Crash of 1929 and the Great Depression

The Great Depression was heralded by the Crash of 1929. It lasted until the onset of the Second World War. In the United States, the effects were grim: industrial production declined by 47%, real GDP by 30%, and unemployment topped 20%[134]. There was a stock market crash, a credit crunch and a banking crisis. As with the twenty-first century's first Credit Crunch, the Great Depression was preceded by cheap credit, a property boom, increasing consumer debt, and rising equity prices[135]. We will begin with these events: the booming Twenties.

### 5.1.1   A Warm Weather Real Estate Bubble

Florida in the mid 1920s – like Florida in the early 2000s – enjoyed steeply rising property prices. Increasing numbers of people were moving to the state, and speculative development boomed to exploit them. By 1926, the supply of new property outpaced the supply of new buyers, and prices started to level out. Then, in the autumn of that year, two hurricanes devastated coastal areas. Investors realised that Florida was not paradise. Prices fell, speculative developments and leveraged investors[136] failed. Mortgage defaults rose dramatically. This in turn caused severe distress to mortgage banks[137].

The Florida boom and bust did not kick off the Great Depression. However, it was a warning that speculative bubbles could develop in the U.S. economy of the 1920s, and that the bust, when it happened, could be both unexpected and severe.

### 5.1.2   Cheap Money and Its Consequences

The run up to the 1929 Equity Market Crash was characterised by easy monetary policy. The FED cut rates in 1927, and borrowing was cheap until the spring of 1928, when it became only a little more expensive[138]. Low interest rates made borrowing attractive, so both corporate and financial leverage increased.

Cheap money alone is not enough to create a boom[139]. In addition you need both a ready supply of money for investment and a broad sense of confidence. The early and mid-1920s were a period of growth and stability. Equity prices were rising, and the market became more than just the talk of a small number of financial experts. Many ordinary people wanted a piece of the action[140]. A number of *investment trusts* were launched: at first, these vehicles offered retail investors the chance to make a diversified equity investment. Thus, rather than buying shares in twenty or thirty companies – roughly the minimum at the time thought necessary to have some measure of diversification – the investor could buy shares in one investment trust, and the trust would in turn invest in a collection of different shares.

The investment trust concept by itself was reasonable. But soon trusts started to use leverage: they borrowed, so that they could invest in more shares. They also bought shares in other leveraged investment trusts, increasing leverage further[141].

Finally, buying *on margin* became commonplace. Investors would use their stock as collateral for a loan, allowing them to buy three, four or sometimes even more times the notional they had to invest. Total margin lending rose as the market did, increasing roughly five fold from the early 1920s to 1928[142]. As prices raced ahead, more and more investors were drawn into the market, and the volumes of shares traded increased fast.

The share price rises of the 1920s were spectacular. The Dow Jones Industrial Average was at 158.75 at the start of 1926, but it had more than doubled, to 381.17, by late 1929[143]. There was a general sense that the ordinary man could get rich quick by investing in stocks.

### 5.1.3 The Crash

When the crash came, the falls were enormous. The Dow fell by more than ten percent two days in a row, October 24th and 25th 1929.

At this point leverage came home to bite. Many investors could not meet margin calls, and were forced to sell. The value of the leveraged investment trusts fell fast and many were forced to liquidate to avoid defaulting on their debt. Those that did not move quickly enough endangered those that had lent them money.

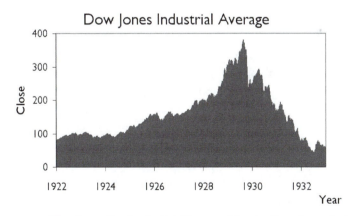

**Figure 5.1**: The Great Crash of 1929 illustrated by the Dow Jones Industrial Average

The market continued in a pattern of small recoveries followed by larger falls throughout the early 1930s, as the figure illustrates. For a few years at least, Americans had learnt that getting rich quick by investing in the capital markets was by no means an assured strategy.

### 5.1.4    The Great Depression

While the 1929 Crash heralded the start of the Great Depression, many economists believe that it did not cause it. Rather they see both the Crash and the Depression that followed it as symptoms of more fundamental economic problems. The macroeconomic debate is beyond the scope of this book[144], but we can identify a familiar pattern in the phenomena[145]:

- Cheap credit, rising asset prices and high investor confidence combine to produce increased investment, particularly increased *leveraged* investment;

- Something causes asset prices to fall;

- Investors sell or – in the case of some leveraged investors who cannot meet margin calls – have their investments sold for them;

- Confidence is wrecked, and borrowing becomes more expensive and more difficult;

- Leverage providers including banks suffer losses and some fail;

- This affects the supply of credit to the real economy, and a recession results.

## 5.2    Political Reactions

The initial government response to the Great Crash was complacent and ineffectual[146]. Many influential bankers at the time thought that the financial system would mend itself. There were attempts to balance the Federal Budget, and frequent exhortations on the virtues of thrift and self-reliance[147]. Monetary policy was tight. This, together with falling economic activity, caused distress to companies. Default rates on loans and corporate bonds rose rapidly[148].

Wages were cut and unemployment surged. Consumer spending declined rapidly, partly due to wages, but also because income inequality had increased in the boom years, and the wealthy can moderate their spending much more than the poor can[149]. It was only when severe unemployment threatened social upheaval that President Hoover began to act[150]. Still, Hoover's responses were neither timely nor adequate. Unemployment rose further, and a banking crisis developed.

It was inevitable that a president who presided over an acute economic depression, massive unemployment, and a failing financial system would fail to be re-elected. So it was that Franklin Roosevelt came to power in the election of 1932. It fell to his administration to try to revive the American economy.

### 5.2.1  Bad Behaviour Broadcast

In the dying days of the Hoover administration, Republicans reluctantly agreed to two initiatives which would later become crucial. The first were hearings under the auspices of the Senate Banking and Currency Committee to investigate the 1929 Crash. These are sometimes known as the Pecora Hearings after the most effective Chief Counsel to the committee, Ferdinand Pecora[151]. What they uncovered shattered the moral authority of the financial community. There were three main strands to the bad behaviour uncovered by these hearings:

- *Crony capitalism and insider trading.* Shares had been offered to friends at well below market prices, company officers had awarded themselves low or no interest loans, and insider information had frequently been used for profit.

- *Fraud, or quasi-fraudulent lack of disclosure.* There were examples of simple fraud – falsified accounts for instance – but also more subtle deceptions. For instance bonds were sold as high quality debt even though the sponsor had reason to believe the issuer would default. This deceit was presumably enacted so that the underwriter could collect their fees.

- *Antisocial behaviour.* To pick one example from many, it was revealed that JP Morgan – one of the richest financiers in America – paid no tax in 1930, 1931 or 1932, perfectly legally. With millions starving, disclosures like this did nothing to enhance the prestige of finance.

### 5.2.2  The Reconstruction Finance Corporation

The other 1932 innovation was the *Reconstruction Finance Corporation* or RFC. This was an independent agency of the United States government, designed to assist in the revival of the U.S. financial system. At first it simply lent to banks against collateral: partly this was necessary as, at the time, many U.S. banks were not members of the Federal Reserve system, and hence the FED could not lend to them[152]. The RFC was also hampered by a requirement that it publish the names of those it lent to, making institutions reluctant to declare publically their need for RFC funds.

A banking crisis developed in Michigan in the dog days between the Hoover and Roosevelt administrations. This quickly spread to neighbouring states, and the RFC seemed powerless to help. The banks did not want to use it. Confidence in the banking system evaporated. Depositors feared for the safety of their money, and many rushed to get their cash out of banks. A nationwide bank run began. By Roosevelt's inauguration day, 4th March, all the states had either declared a bank holiday or restricted the withdrawal of deposits. Roosevelt extended this holiday period until the 13th March[153]. The closing of the banks provided a short pause, but quick action was needed to prevent widespread bank failures once the system reopened.

## 5.3   The New Deal

The responses of the Roosevelt administration to the economic problems of the Great Depression are collectively known as *The New Deal*. There were a large number of programs: these are now regarded as ideologically incoherent and of varying efficacy. This is partly because the stock market crash of 1929 was so sudden and so severe that there was a strong desire to do something, but many competing ideas of what that thing might be.

FDR's men first had to keep the banking system from complete collapse. Their approach is familiar from the previous chapter: the FED was instructed to swap banks' assets for cash. This infusion of liquidity gave the banking system some breathing space. Meanwhile, the mandate of the RFC was extended so that it could not just lend banks money, but also buy capital instruments[154]. Recapitalisation of the banking system proceeded alongside a much freer supply of money. It is worth noting that the authorities expected the banks, in turn, to do their bit for the economy. Jesse Jones, chairman of the RFC, told the American Banker's Association that 'Banks must provide cash ... otherwise the government will.' A credit crunch was not going to be permitted to exacerbate the damage to the American economy.

Given that the 1929 crash marked the start of the Great Depression, there was an association of stock market falls and economic calamity in the public imagination. This created a sense that something must be done to prevent the equity market from disrupting the U.S. economy to such a large extent in the future. Furthermore there was a need to deal with the unsavoury behaviour in the securities markets uncovered by the Pecora Hearings. Two programmes of action were therefore needed:

- Something to prevent dislocations in the securities markets from damaging the banking system; and

- Something to make the securities markets more honest.

Thus the New Deal reforms included the Banking Acts of 1933-35 and the Securities Acts of 1933-34.

### 5.3.1 The '33 Act and Securities Regulation

Two securities acts – the Federal Securities Act of 1933 and the Securities Exchange Act of 1934 – drastically changed the regulation of securities in America. Together they created the *Securities Exchange Commission* or SEC to oversee the securities markets; required securities issuers to disclose all material information to securities buyers; and made fraudulent or misleading disclosures illegal. In particular, issuers of securities are required to register them with the SEC, and to provide the SEC with audited financial statements, and with detailed information on the equities or bonds to be issued. This meant that all companies that desired access to American capital markets must give the SEC regular information on their financial condition[155], a resource that has proven very useful to investors ever since.

### 5.3.2 Banking and the Glass-Steagall Acts

The Great Depression took a massive toll on U.S. banks. Over ten thousand failed or had to merge, and the total number of banks fell by over a third. Clearly something needed to be done to restore the public's confidence in the financial system. Therefore the Banking Act of 1933 established wide-ranging *deposit protection*[156]. Retail deposits – providing they are not too big – are protected by a Federal Government Agency, the *Federal Deposit Insurance Corporation* or FDIC. If the bank fails, the FDIC will take it over: depositors' funds are not at risk. This allows retail investors to leave money with a bank without worrying about its financial condition. In particular if there are rumours about the possible failure of a bank, then retail depositors will not lose and so they have no incentive to take their funds out. This helps to prevent a *bank run* where the rumour of a bank failure causes deposits to be withdrawn and hence makes the rumour true[157]. (Bank runs are so called because in the days before deposit protection people would literally run to the bank to get their funds out before it failed.)

Securities were the most visible culprit for the nation's financial distress. Therefore it is hardly surprising that some politicians wanted to 'protect' the banking system from the dangers of securities trading, and so to allow it to function better should the equity market suffer another crash. The Pecora Hearings gave those politicians who had always distrusted the securities markets – including a certain Senator Carter Glass – the excuse they needed to separate banking and securities dealing. Clauses were added to the 1933 Banking Act to cleave the two asunder[158]. In particular – at least after a second bite of the cherry was taken in the Banking Act of 1935 – there were four main changes:

- Banks were forbidden from purchasing a significant amount of U.S. securities for their own account, preventing them from speculating in the securities markets with depositors' money;

- They could only purchase and sell securities when they had a direct customer order;

- Securities issuance – including underwriting and selling equities or corporate bonds – were forbidden to banks[159]; and

- Banks could not be affiliated with entities engaged principally in securities underwriting, distribution, or dealing.

These provisions are collectively known as the *Glass-Steagall Act* as Senator Glass was one of the main driving forces and Henry Steagall, as Chairman of the House Banking and Currency Committee, helped him get the legislation passed. Together, the Glass-Steagall provisions reduced the cost of deposit protection to the U.S. government by making banks concentrate on banking rather than securities activity. But they also split American financial services in two. If you wanted to trade and underwrite securities, you could not take deposits and make loans. Without Glass-Steagall, American banks would have continued to be active in securities trading, as their European peers were. But post Glass-Steagall there was an opportunity for specialist securities firms to grow up: the *broker/dealers*. These firms are not banks: they are regulated by the SEC rather than banking regulators. By the 1990s, the largest five of them were dominant players in the Capital Markets[160].

It is worth noting that the bankers fought reform firmly and consistently. Without the loss of their moral authority after the Pecora Hearings, and the Democratic party's domination of the White House, the Senate and Congress, the Banking Acts would probably not have been possible. Government was now treating finance as 'an erratic and irresponsible force requiring strict social discipline'[161].

### 5.3.3 *Lending in a Depression*

When a bank loses money, it can lend less. To see this, consider the simple old-style bank with £8M of capital we met in the previous chapter. This capital allows it to take risk[162].

The bank's regulators force it to keep at least 8% of capital against loans. This means that the bank can lend up to £100M. To be prudent it lends only £90M and takes £82M of deposits to fund the lending. Figure 5.2 illustrates the situation. Here we have used the accountant's term 'shareholders' funds' for capital.

| Assets | Liabilities | |
|---|---|---|
| Loans   90 | Shareholders' funds | 8 |
| | Deposits | 82 |

**Figure 5.2**: A very simple old style bank

Now suppose that defaults rise in the bank's loan book and it is forced to write off £1M of loans. The problem is that this loss must be deducted from capital. This reduces the bank's capital to £7M and on this basis it can only lend $^7/_{.08}$ or £87.5M. That is less than its actual lending. The bank needs more capital, and it certainly cannot make any more loans until it has it.

Losses, then, restrict a bank's ability to lend. Another problem – even without the regulatory restriction of 8% of capital[163] – is that banks need to borrow to fund their lending. They do this either by taking deposits, as in our example, or by borrowing on the interbank market. But in a depression, retail depositors are suffering and do not have as much money in their bank accounts. Moreover interbank borrowing is harder because banks distrust each other's credit quality. Term borrowing becomes particularly expensive as the banks' credit spreads blows out, but banks' tolerance of liquidity risk decreases, forcing them to seek out term borrowing for their liquidity needs even as it gets more costly.

Banks are reluctant to lend to borrowers of uncertain or declining credit quality. Loans which in times of confidence would have been made without a second thought are declined or granted only under much more onerous interest.

Finally, there is confidence. In a crisis, banks like to make positive communications with their shareholders. 'We have stopped this form of risky lending,' is viewed as a prudent message. That the form of lending concerned is profitable and necessary for the borrower is entirely irrelevant. Similarly banks decrease their risk to reassure the markets and to try to maintain confidence[164].

Lower capital, and less easy and cheap funding, uncertain counterparty credit quality, and the desire to be seen to be reducing risk all create a credit crunch.

This type of crunch occurred during the Great Depression. Mortgages became very hard to get and very expensive when they could be obtained. Many home owners struggled with payments on existing mortgages making even those lenders with enough capital and funding reluctant to take on new risk[165]. Foreclosures were causing misery across the country: by 1933 there were a thousand homes being repossessed every day[166].

Distress like this had all sorts of knock-on effects. For instance, the building industry suffered as very few new houses were being built. Something had to be done.

## 5.4 The RFC and Other Rescuers

Roosevelt had few of the scruples of modern central bankers. For instance, he explicitly declared that 'the definitive policy of the government has been to restore commodity price levels'[167]. It was a priority, then, to restore prices to the point where firms could make money, wages could be paid, and debts serviced. Roosevelt's men were going to intervene in the markets, and the RFC was one of their instruments[168].

### 5.4.1  The RFC Becomes Bold

Lending banks money is of little use if they are not solvent. What they need then is capital, and this is what the RFC started to provide in 1933. Legislation was changed so that the RFC could inject capital into the banks[169], and this, together with deposit protection, put the banking system back on a firmer footing. It also made the RFC a significant instrument of state capitalism: by 1935 it owned over a billion dollars of capital (roughly two hundred billion in 2007 dollars[170]) in over half the banks in the country.

As the thirties progressed, the importance of the RFC's rôle in the economy increased. By 1938 it had disbursed over $10B. It was the largest single investor in the American economy. On the basis of its lending, it was the largest bank in the country.

### 5.4.2  Supporting Home Ownership

There are few things more damaging than losing your home. And if a high foreclosure rate is bad for society, it does not help the banks either: as we saw in chapter 2, once bank sales of foreclosed homes start to dominate the property market, liquidity evaporates, and prices plummet. Mortgage collateral – houses – can no longer be sold for more than the loan they secure, causing losses for foreclosing lenders. President Hoover had already tried to address the problem, but his Home Loan Bank Act of 1932 had done little. Roosevelt wanted to do more to protect home owners from foreclosure.

The Home Owners' Loan Corporation or HOLC was the result[171]. Beginning in 1933, it took over unaffordable mortgages and restructured them, extending the mortgage term. The HOLC's government status allowed it to borrow cheaply. These savings were passed on to mortgage borrowers too, resulting in much lower payments on the restructured loans. The HOLC prevented the collapse of the property market. Like the RFC, it grew furiously, eventually having a rôle in roughly 20% of all non-farm mortgages.

### 5.4.3 The Taxpayer Takes Default Risk

The next step was to persuade American banks to get back into the lending business. One of their concerns was clearly the plunging credit quality of individuals during the depression. So the government set up the *Federal Housing Administration* or FHA to insure against the risk of default on mortgage loans. The ability to buy protection on mortgage default from the government encouraged lenders to lend. For the first time they could hedge the risk of non-performance of mortgages. In particular it meant that lenders could be sure of a profit on an FHA-insured loan: provided the interest on the mortgage was more than the cost of insurance, the cost of funding the loan, and the cost of servicing it, they would make money with little risk.

The FHA helped in giving lenders a way of reducing the risk of mortgage lending, but it did nothing to assist in funding. A lender still had to raise money if it wanted to make new loans. The answer to this problem came with the foundation of the Federal National Mortgage Association in 1934. This entity, which rapidly became known as *Fannie Mae*, was originally part of the RFC[172]. It purchased FHA-insured mortgages, freeing up cash, and so allowing the lender to lend again.

Thus between the RFC, the HOLC, the FHA, and Fannie, a series of substantial interventions recapitalised and reliquified the banking system, supported property prices, and encouraged banks not to limit their involvement in mortgages to foreclosing on existing customers.

### 5.4.4 The Secondary Mortgage Market

Fannie created a *secondary market* for residential mortgages. Lenders no longer had to keep all the loans they made. Provided Fannie would buy them, their loans could be turned into cash, giving the primary lenders the confidence to keep lending whatever the economic climate. In this sense the Roosevelt administration's reforms achieved their aim. However there was a price. First, Fannie's borrowings counted as government debt and thus they inflated the size of government liabilities. And second, the mortgage market split into those loans that were eligible for sale to Fannie, known as *conforming* loans, and the rest.

### 5.4.5  *Lyndon Johnson and the Strange Status of Fannie*

The first thirty years of Fannie's life were fairly orthodox. It bought FHA-insured mortgages, funding them with its borrowing. Its functions expanded gradually: in 1944 Fannie was permitted to buy mortgages insured by the Veterans' Association, giving it a rôle in ensuring that ex-Service personnel could get a home loan. It was also allowed to provide 'special assistance' to the housing market, buying uninsured mortgages if this was necessary to prevent a decline in home building.

Fannie was in some ways too good at its job. By 1968, it had a significant portfolio of mortgages, and significant borrowing to fund them. The President at the time – Lyndon Johnson – was worried about government finances, especially as the costs of the Vietnam War were escalating. In a move that has echoes of the creative accounting used by Enron more than forty years later, a structure was developed that changed the appearance of the U.S. government finances without a significant change in their substance.

The idea was a peculiar kind of semi-privatisation which removed Fannie's debts from the roll call of government liabilities. Fannie was split: a new entity, Ginnie Mae, took over responsibility for buying FHA- and VA-insured loans; while Fannie itself was left to support the broader secondary mortgage market by buying loans directly from the primary lenders. At the same time Fannie was listed and shares were sold to investors.

However, unlike a true privatisation, the government hinted that Fannie's liabilities would continue to be backed by the U.S. Treasury. President Johnson could not explicitly guarantee Fannie: that would not have removed the firm's liabilities from the official calculation of government finances. But he could give a nod and a wink to the market. This ambiguity is part of the untidy border between government and private enterprise that characterises American capitalism.

Clearly a quasi-private Fannie would enjoy a monopoly position in the secondary market. Therefore a competitor was set up to keep it honest: *Freddie Mac*[173].

Both agencies were permitted to buy conforming mortgages in the secondary market. The status of Fannie and Freddie is acknowledged in the term used to refer to them and a number of other similar agencies acting for the government: *government sponsored enterprise* or GSE. We will call them F&F.

F&F have a variety of other advantages in addition to implicit support from the government: they have their own regulator, the Office of Federal Housing Enterprise Oversight[174], various tax breaks, and even an exemption from SEC registration of their securities.

### 5.4.6   Private Gain, Social Loss: Subsidising Fannie and Freddie

The problem with the structure Johnson ended up with is that if things go well for F&F, then their shareholders benefit, but if things go badly, the taxpayer may end up footing the bill. Specifically if Fannie or Freddie (or both) buy too many mortgages which go on to default, their capital will be eroded. If this happens too fast or too far, their regulator will be forced to step in. Even before that, losses may make investors unwilling to buy the GSE's debt – despite the government guarantee – and this would pose a real threat to their business model. They cannot continue to buy mortgages without new borrowing. The government would be forced to intervene and make good on Lyndon Johnson's implicit promise.

There is nothing like the GSEs in any other major market. Government agencies are common around the world. Private corporations are ubiquitous. But enormous enterprises with shareholders but whose debt bears the strong suggestion (but not an absolutely binding legal promise) of a government guarantee are unique to the United States. This 'implicit guarantee' is valuable – one estimate in 2005 put the taxpayer subsidy to F&F at around $150B, of which Fannie and Freddie's shareholders have taken roughly half[175] – and the GSE structure does not clearly offer the best way for the U.S. taxpayer to support the mortgage market. The price of Lyndon Johnson's accounting sleight of hand has been a considerable transfer of wealth to Fannie and Freddie's shareholders and management.

### 5.4.7   Conforming Loans

In order to protect the taxpayer, F&F are only permitted to buy reasonably safe loans. The definition of *conforming loan* has changed significantly over the years, but it has three key features:

- A *size limit*. F&F were supposed to be helping the ordinary American afford housing, not the ordinary film star or basketball player, so loans that are too big are not conforming. The size limit has been raised over the years as house prices have gone up. Loans that are bigger than the conforming limit are known as *jumbos*.

- *Documentation requirements.* In particular stated income loans and related low documentation structures are not conforming.

- An *80% LTV limit.* Loans beyond this have to have private mortgage insurance.

### 5.4.8    Yet Another Institution: The FHLBs

Hoover's 1932 Home Loan Bank Act created one final piece of financial system architecture that will be important later, the *Federal Home Loan Banks* or FHLBs. The FHLB system was a kind of mini mortgage-based Federal Reserve system. Like the FED, there are twelve regional banks (although the FHLB regions are different from the FED districts): these regional FHLBs lent against collateral, but mortgage collateral rather than bonds which are eligible at the FED window. The whole system was government guaranteed, so the FHLBs could borrow cheaply[176].

Much of FHLB lending is to the thrifts[177]: the original aim was to liquefy mortgage assets within the banking system and hence make mortgages more affordable. Unlike the Federal Reserve system, the FHLBs are owned by the banks they support, so their equity is held by over 8,000 financial institutions in the U.S., and is not publically traded.

### 5.4.9    A Fractured Landscape

The New Deal reforms enhanced the safety and efficiency of the American financial system. Deposit protection removed the risk of losing their savings from many Americans. The combination of the HOLC, the FHA, Fannie, and the FHLBs made mortgages cheaper. Long term level pay structures slowly replaced the short term interest only loans that had predominated before the Great Depression[178]. Home ownership became part of the American Dream. Glass-Steagall separated securities trading and underwriting from banking, enhancing the safety of the banking system (at least for a while).

However, the resulting system was enormously complicated. There were lots of different types of institutions: depositary banks and broker/dealers; thrifts and FHLBs; the FHA and Fannie. Different regulators abounded too: the Federal Reserve and the OCC for banks, plus the FDIC; the SEC for the broker/dealers; a separate regulator for the thrifts; what became the Federal Housing Finance Board for the FHLBs; and the Office of Federal Housing Enterprise Oversight for Fannie and Freddie. Each regulator had different rules, different concerns, and a jealous concern for the health of its charges. It is perhaps inevitable that some regulators judge their importance by the size and status of the firms they supervise. And a regulator suggesting that the whole class of firms under their charge are dangerous is akin to a turkey voting for Christmas. This diversity of regulators may therefore contribute to financial instability: they certainly make financial system reform more difficult.

## 5.5   The Evolution of Freddie and Fannie

By 2000, F&F bought mortgages in the secondary market; kept some of them, managing the risks of default and of prepayment; and issued securities to pay for those mortgages. Some of these bonds have a structure which allows the GSEs to pass on the prepayment risk of mortgages to the bond buyer. They are a kind of ABS – actually the first kind of ABS to be issued – so we begin by introducing them.

### 5.5.1   The Use of Mortgage Backed Securities

If Fannie and Freddie only guaranteed mortgages – like the FHA – then their risk management would be relatively simple. All they would have to do would be to charge a sufficient fee for their guarantee to cover the likely number of mortgage defaults, plus a bit more to provide a return to shareholders.

However Fannie and Freddie buy mortgages, sometimes keeping them for themselves rather than selling them on into the secondary market. They get the cash to do this by issuing bonds. At first Fannie and Freddie issued standard term bonds, similar to a government bond. Thus for instance Fannie might issue a ten year bond paying five percent a year: the proceeds of this were used to buy mortgages yielding seven or eight percent, giving a profit for the Agency.

Thanks to the market's belief that the GSEs had government support Fannie and Freddie were able to raise money cheaply. However it quickly became apparent that they had a problem: how long should they raise money for? This question arises because of the prepayment risk in mortgages.

- If they issue long term bonds and interest rates fall, lots of mortgages prepay, and they have to pay interest on their bonds for a long period. Their mortgage assets disappear and they cannot find new ones to replace them that pay as much (since rates have fallen);

- But if they issue short term bonds and interest rates rise, then the mortgages will not prepay nearly as fast, and they will have to refinance their debt more expensively.

The Federal Agencies solved this problem using a kind of ABS[179]. The ABS structure we described in the last chapter was not quite right for the Agencies: in that version, the originator sold the mortgages to the SPV, so the SPV had both the prepayment risk of the mortgages and the default risk. ABS investors had some protection from default thanks to the SPV's equity, but not complete protection. The Agencies wanted to pass on all the prepayment risk but they wanted to keep the default risk: that was their job, and bonds without default risk would be more attractive to investors, especially when times were tough in the mortgage market. Therefore they *guaranteed* the performance of the mortgages backing their bonds: if any of them defaulted, the agencies undertook to make good the missing payments.

This form of ABS is called a *pass through*. Monthly mortgage payments are collected by the originator and split three ways:

- The originator takes a fee for servicing the mortgage;

- They pass the rest on to the Agency. The Agency takes a fee for their guarantee, and undertake to pay the ABS holder the mortgage principal if the loan defaults;

- The Agency then passes the remaining cash on to the ABS holder, along with any sums they are due because of defaults.

The ABS holder thus receives most of the interest and all of the principal payments on the underlying mortgage pool. If rates fall, their bond will repay most of its principal in a few years; whereas if rates rise, the bond might still be repaying significant amounts of principal after twenty or more years. Pass through buyers are paid to take this risk, while Fannie and Freddie solved their prepayment risk management problem. Now instead of trying to guess the right duration for their unsecured debt, they could sell ABS backed by portfolios of mortgages. These *mortgage backed securities* or MBS had more predictable behaviour than individual mortgages. They were attractive to bond buyers without expertise in direct mortgage lending.

### 5.5.2   Pass Throughs: Who Got What?

The issuance of MBS allowed F&F to buy more mortgages without taking too much prepayment risk[180]. Fannie and Freddie guaranteed payments on the MBS, so although prepayment meant investors did not know *when* they were going to get paid precisely, they did have confidence that they were going to get paid eventually. This attracted investors who did not want to take default risk, expanding the pool of funds available to support primary mortgage lending.

| Initial Lender | Funding | Prepayment risk | Default risk |
|---|---|---|---|
| Retains Loan | Lender | Lender | Lender |
| Sells to Aggregator | Aggregator | Aggregator | Aggregator |
| Sells to GSE, Loan Repackaged into MBS | MBS Buyer | MBS Buyer | GSE |

**Figure 5.3**: The holders of various risks for a conforming mortgage depending on the strategy of the initial lender

### 5.5.3 The Growth of Private Label MBS

Mortgage Backed Securities issued by the GSEs soon became an important part of the bond market. Investors became comfortable with the prepayment risk of MBS especially as they were paid a considerable spread over U.S. treasuries for buying Freddie- or Fannie-issued MBS.

Once this market had established itself it was natural for the large mortgage originators to test investors' willingness to take default risk as well as prepayment risk. After all, why should the GSEs have all the fun? Thus in 1977 the first true *private label* MBS was issued by Bank of America. Like a GSE pass through, this passed through all the payments on a fixed pool of mortgages to the investor. Unlike a GSE deal, there was no guarantee, so the investor suffered if the default rate for the mortgages in the pool was higher than expected. To combat this, the first private label MBS deals were backed by very high quality mortgages: it took some years for investors to become comfortable with lower quality pools.

Two big boosts for the private label MBS market came in the 1980s. In 1984 regulated financial institutions were allowed to buy private label MBS; then in 1986 changes to the tax code made it possible for more sophisticated types of MBS to be issued.

Private label MBS issuance developed fast after that, and by 2005 roughly the same volume of MBS was issued by the private sector as by the GSEs. Originators could now pass on whatever mortgage risks they did not like. The key technology needed for an intense real estate crunch was in place.

# Notes

[134] For an old but still captivating account of the Great Depression, see J.K. Galbraith's *The Great Crash 1929*, Penguin Books.

[135] See *The Great Depression as a credit boom gone wrong*, by Barry Eichengreen and Kris Mitchener, BIS Working Papers No. 137 available at www.bis.org/publ/work137.pdf.

[136] One of the features of the 1920s Florida boom was the trading of *binders*. These were the right to buy a house at a fixed price in the future for a small down payment, so a derivatives trader would call them futures. The small down payment relative to the value of the property meant that investors in binders had considerable leverage.

[137] For more on the Florida property price boom and bust, see *A Brief Florida Real Estate History* by J. Bruce Cumming, Jr., available at www.clas.ufl.edu/users/thrall/class/g3602/floridarealestatehistory.pdf, while for the knock-on impact on the state's banking system, see *Panic in Paradise, Florida's Banking Crisis of 1926*, by Raymond Vickers (1994).

[138] The FED was not united on the cheap money policy even at the time. After the fact, Adolph Miller of the FED reserve board called it 'one of the most costly errors' committed by the FED in the last 75 years.

[139] This point is Galbraith's.

[140] A famous financier – the tale is attributed variously to Joe Kennedy or J P Morgan – is reputed to have sold most of his holdings just before the 1929 crash because he was given a stock tip by his shoe shine boy. Apocryphal or otherwise, the story illustrates an important insight: if people who do not work in finance start to talk about the prices of financial assets, an unsustainable boom may well be in progress. The growing acceptability of making dinner party conversation about property prices in the 2000s was a similar signal.

[141] The leveraged investment trusts of the 1920s were similar to the SIVs of the early to mid 2000s. SIVs invested in credit instruments and investment trusts in equity, but otherwise the similarities are striking.

[142] When rates were raised in 1928, it was not enough to curb margin loans, but it was enough to cause damage to the real economy.

[143] For Dow Jones data going back to 1896, see www.djindexes.com.

[144] This is a good thing: Bernanke calls such an understanding 'the Holy Grail' of modern macroeconomics. See Ben Bernanke et al., *Essays on the Great Depression*, 2000 for more details.

[145] This is a (crude) version of Irving Fisher's Debt Deflation account of busts which in turn informed the work of Hyman Minsky.

[146] For instance President Hoover opined in 1930 that 'we have now passed the worst'. The worst was still two or three years off.

[147] For much, much more on the background to the Great Depression and the policy responses it produced, see Volumes 1 (*The Crisis of the Old Order*) and 2 (*The Coming of the New Deal*) of Arthur Schlesinger's *The Age of Roosevelt*, 1951-1953.

[148] Note that bond repayments are fixed – 'given in nominal terms' in the language of macroeconomics – so if it is harder to earn a dollar, it is harder to repay a dollar.

[149] Thus an economy with high income inequality is also more prone to severe contractions in consumer spending.

[150] There was also a tension between the right to private enterprise thought to be guaranteed under 14th Amendment protections of liberty and property, and the government's power to intervene in the economy. Schlesinger quotes Judge Brandeis, dissenting in the case of New State Ice Co. vs. Liebmann (285 U.S. 262) as follows: 'There must be power in the states and the nation to remould, through experimentation, our economic practices and institutions to meet changing social and economic needs.' It was only the length and intensity of the Great Depression that overcame the reticence to intervene and so made the New Deal possible.

[151] See *Stock Exchange Practices - Introduction to the Pecora Committee Report*, 73rd Congress Report No. 1455, available at www.sechistorical.org/collection/papers/1930/1934 _06_06_Intro_to_Pecora_C.pdf.

[152] At the time, national banks were required to be members of the FED system but state-chartered banks were not. The U.S. banking system had many small banks – over 20,000 in 1929 – and these were highly vulnerable. Of the 3,635 bank failures in 1930-31, over 85% were in institutions with a capitalisation of less than $100,000. See *Fifty Billion Dollars* by Jesse Jones, 1951.

[153] Jones' book, previously cited, gives more details of this, and of RFC interventions.

[154] Most of the RFC's support actually took place via the purchase of preferred stock rather than equity.

[155] For examples of SEC reporting see www.sec.gov/edgar/searchedgar/webusers. htm.

[156] Or at least it did once the state banks were corralled into the system in 1936. Roosevelt opposed deposit protection at first, only acquiescing after it was added to the '33 Act during its legislative passage by Senator Vandenberg.

[157] Bank runs had been a regular feature of financial crises before the advent of deposit insurance. See for instance Charles Calomiris's *Bank Failures in Theory and History: The Great Depression and Other "Contagious" Events*, National Bureau of Economic Research Working Paper 2007.

[158] For more details, see *The Crash and Its Aftermath: A History of Securities Markets in the United States 1929-1933* by Barrie Wigmore.

[159] The exceptions to this included government bonds, certain state and other municipal bonds, and some activities relating to commercial paper. The main aim of this provision was to ensure that unsound loans were not made to 'buy' securities business and that there was no conflict of interest between lending money to a securities issuer and marketing its securities.

[160] In retrospect it is surprising that a more radical reform of American financial services was not attempted. After all, the banking system of the twenties had failed society badly and it needed massive amounts of Federal money to survive. Perhaps one reason for the comparative moderation of the New Deal reforms is simply that the only people who knew enough about the banking system to propose credible changes were orthodox economists.

[161] The quotation is from Schlesinger, Volume 2.

[162] So, at least to a first approximation, capital is either money the bank has got for issuing shares or profits it has made but not distributed to shareholders.

[163] Capital requirements were not an issue in the Great Depression, but they are a significant factor in the Credit Crunch.

[164] Thus, according to the Federal Reserve *Banking and Monetary Statistics*, we see loans as a percentage of bank deposits falling from 80% in the late 1920s to around 50% in the 1930s.

[165] In the 1920s and 30s the typical loan made for the purchase of a home was much shorter term than today, perhaps as little as three or five years. This meant that borrowers had to refinance their homes frequently: if this was not possible or not affordable then they lost their homes to foreclosure. LTVs were also much lower, requiring house borrowers to have a much larger deposit than today and so making it much more difficult to buy a property. As a result less than half of American households owned their own home.

[166] Further details can be found in *The Federal Response to Home Mortgage Distress: Lessons from the Great Depression* by David Wheelock, Federal Reserve Bank of St. Louis Review, May 2008. This is available at `research.stlouisfed.org/publications/review/08/05/Wheelock.pdf`.

[167] *Public Papers and Addresses* (II, 138), quoted in *The Economic Thought of Franklin D. Roosevelt and the Origins of the New Deal* by Daniel Fusfeld, 1955.

[168] For the other mechanisms of the Fair Deal interventions, see Schlesinger and Fusfeld.

[169] Typically the RFC bought preferred stock rather than equity. Their approach, according to Jones, was first to save those banks which were nearly solvent (defined as assets worth 90% or more of liabilities). It was important to act speedily on these 'nearly there' cases since only solvent institutions qualified for FDIC insurance. Other patients were put into longer term care: some of them made it, some did not.

[170] The business of adjusting for growth over long periods is delicate. Wages grow faster than the cost of goods so over a long period, inflation adjustment may under-estimate the value of a past sum today. These complications mean that there is considerable debate about the 'correct' calculation of inflation, with many arguing that the best approach is to use a wide measure like the GDP deflator.

[171] Wheelock, referred to earlier, is especially good on the Home Owners' Loan Corporation.

[172] Fannie arose, according to Jones, because private capital was reluctant to invest in National Mortgage Associations. These had been provided for in the National Housing Act of 1934, but banks did not take up the RFC's offer of dollar for dollar capital to invest in them. No one wanted to be in the lending business. So, with typical brio, Jones did the job himself, and set up both a RFC Mortgage Corporation to lend directly, and Fannie. Additional information on Fannie can be found at `www.fanniemae.com`.

[173] Freddie was originally the Federal Home Loan Mortgage Corporation. Its charter outlining the rôle of the agency can be found at `www.freddiemac.com/governance/pdf/charter.pdf`. Note too that another (perhaps more important) motivation for the creation of Freddie was the rise in U.S. residential mortgage costs during the late 1960s. Fannie could not buy enough mortgages to keep home finance costs down: Freddie was created to provide more funding for primary mortgage lenders. The belief was that the resulting saving in the cost of funds for lenders would be passed on, reducing costs for borrowers.

[174] See `www.ofheo.gov` for more details.

[175] See *The GSE Implicit Subsidy and the Value of Government Ambiguity* by Wayne Passmore, Federal Reserve Board Finance and Economics Discussion Series 2005-05 available at `www.federalreserve.gov/Pubs/feds/2005/200505/200505pap.pdf`. Passmore's estimate for the implicit government subsidy of Fannie and Freddie is between $122B and $182B, with shareholders retaining between $56B and $106B.

[176] See `www.fhlbanks.com` for more details.

[177]The thrifts, or *savings and loan associations* started life as mutually held banks: they were owned by their depositors and borrowers. Nowadays some of the thrifts have demutualised, but their lending remains concentrated in the mortgage market.

[178]The typical loan type in the 1920s was a three or five year interest only structure, strongly reminiscent of a 2/28 or 3/27 teaser ARM.

[179]The first ABS were actually GSE-issued bonds. Ginnie Mae was the first issuer, in 1970, with Fannie following soon afterwards.

[180]They still had *a lot* of risk if interest rates changed, just not so much that it threatened their solvency. See *The Interest Rate Risk of Fannie Mae and Freddie Mac* by Dwight Jaffee, Journal of Financial Services Research 2003, available at `faculty.haas.berkeley.edu/JAFFEE/ Papers/FinalJFSR.pdf`.

# Chapter 6

---

## Securitisation, Tranching, and Financial Modelling

*It is evident that Adam, with all his science, would never have been able*
*to demonstrate that the course of nature must continue uniformly the same,*
*and that the future must be conformable to the past.*
David Hume

---

## Introduction

The Credit Crunch has been referred to as a STD: a Securitisation Transmitted Disease[181]. There may be an element of truth in that characterisation, but it ignores the fundamental cause of the malaise: bad mortgage lending.

Still, without securitisation, bad mortgages would only have affected lenders, so it is worth looking at the technology, what it can do, and why it is used. In particular the securitisation of subprime mortgages is discussed to give some insight into why the prices of ABS backed by that particular type of collateral fell so fast.

A key part of the story of the Crunch concerns model failures. Securities were mis-priced and the risk of them mis-estimated. This happened not just in the Ratings Agencies – who gave the highest AAA rating to securities that a few months later had halved in value – but also in some of the leading investment banks. We look into model risk and show how the assumption that the future will be like the past led to enormously optimistic assessments for subprime-backed ABS.

Some financial institutions believed that these bonds were genuinely low risk. In some cases they bought a lot of AAA-rated subprime ABS. These firms suffered large write-downs when house prices started to fall. Some of the biggest losses are discussed together with the reasons for and consequences of those losses.

# 6.1   Securitisation

The process of creating asset backed securities, described in section 3.1.7, is known as *securitisation*. So far, simple securitisations have been described. Much more complicated structures were created in the ten years prior to the Crunch. What were the building blocks for creating these more complex asset backed securities and how were they used?

### 6.1.1   What Can Be Securitised?

Thus far we have seen ABS where the assets backing the security were residential mortgages. These bonds, sometimes known as *mortgage backed securities* or MBS, are the most important type of ABS[182]. Other assets can be used too, though. Prominent examples include collections of:

- Commercial mortgages;
- Bonds;
- Corporate loans;
- Credit card receiveables.

All of these assets share common characteristics which make them good candidates for a securitisation:

- They are *self-liquidating*. That is, if all goes well, the asset turns into cash. In contrast an asset like a Velázquez has to be sold in order to realise its value;

- Having many individual assets in the collection giving some measure of *diversification*. One bond issued by a company might not be repaid if the company goes bankrupt, but if we have a hundred bonds from a hundred different good quality companies in different industries and countries, then it is highly unlikely we will lose a large fraction of the amount lent, since it is unlikely that lots of these companies will go bust at once;

- The history of comparable assets is a guide to the future. If we know that the charge-off rate on prime credit cards in the last year was 1%, then it is unlikely to be 5% next year (unless something changes);

- Statistics can easily be gathered which give insight into the risk of the portfolio. For instance, for the bonds we might ask what the ratings and credit spreads are, which industries and countries the issuers are in, what the maturity of the bonds is, and so on.

### 6.1.2    Tranching

Suppose we have a diversified portfolio of 100 investment grade corporate bonds, with $10M of each bond. The total loss if every corporate defaults with zero recovery is $1B, but this loss is vanishingly unlikely. If the bonds have a five year maturity, we might expect a handful of defaults over their lives. No defaults at all is possible if we are very lucky: whereas with bad luck we might conceivably see ten or even fifteen defaults.

The average corporate recovery rate is 50%, so with ten average sized defaults over the five year period, the total loss would be $50M. Thus the range of losses we might expect go from zero to a hundred million or so. The loss distribution is as shown in figure 6.1[183].

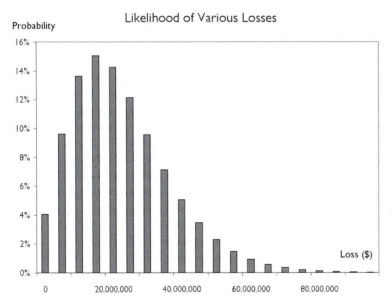

**Figure 6.1**: The probability of various numbers of defaults in a portfolio of 100 loans

Different investors have different attitudes to risk. Some want a low risk investment, and in consequence are happy to have (or at least reluctantly willing to agree to) a low return. Others want a higher return in exchange for taking more risk.

We can satisfy many desires by *tranching* the securities we issue. That is, we take the $1B of securities we need to issue in order to pay for the bonds, and split them into three classes.

The *Junior* or *Equity* tranche will absorb the first losses. Suppose we issue $40M of this. There is only a 4% chance that there will be no defaults, so this security will likely not return its full principal. However if it pays a high enough interest rate then we can perhaps persuade an investor to buy it.

Once the equity is in place and able to absorb losses up to $40M, the next securities are much safer. There is a roughly 90% chance that losses will be less than $40M, so nine times out of ten the equity will absorb the losses due to defaults.

The next step is to issue $50M of a *Mezzanine Tranche*. There is some chance that this will not return its full principal, so we do have to pay a reasonable credit spread to investors in order to persuade them to buy it. On the other hand, it is much less risky than the equity, so the credit spread will be lower than the return equity investors will expect.

The equity and the mezz together absorb $90M of losses. There is only a 0.05% chance of losing more than that, based on our model, so the final $910M of securities are very safe. This *Senior Tranche* might only pay a coupon of some teens of basis points, and it will be very highly rated. Figure 6.2 illustrates the securitisation.

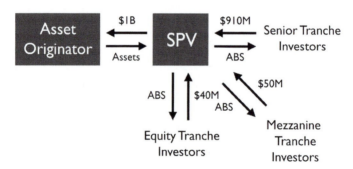

**Figure 6.2**: A simple CDO

Carving the possible losses into pieces and issuing securities which take various parts of the loss distribution allows us to satisfy the needs of various investors: low risk low return players can buy the senior, while those who are hungry for yield can buy the mezz. Of course more than three tranches are possible, and real securitisations have anywhere from two to ten or more tranches.

The structure whereby bonds are used as collateral for the issuance of tranched ABS is known as a *collateralised bond obligation* or CBO. Similarly, a deal with tranched loans is a CLO, and 'CDO' – for collateralised debt obligation – is the generic term. The version for mortgages is sometimes called a CMO, reasonably enough, or less transparently a REMIC (for real estate mortgage investment conduit, a name introduced for the issuing SPV when U.S. law was first changed to make tranched MBS practical in 1986).

### 6.1.3 It's Raining Cash

The structure of the tranches is sometimes known as the *waterfall*. Cash paid by the collateral falls first to the most senior tranche, then on down through the mezzanine to the junior. Suppose the issued securities are scheduled to pay interest every three months. At the start of each quarter, we begin directing cashflows from the collateral into an account. Thus in a CBO whenever one of those bonds pays interest, that payment goes into the account. Once that account has got enough in it to meet the next interest payment on the senior debt, we redirect the flow of cash into an account for the benefit of the mezzanine. When that has whatever it needs, the equity tranche gets any remaining cash we receive in the quarter[184].

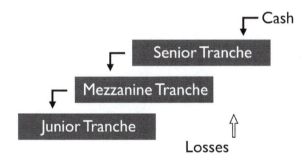

**Figure 6.3**: The waterfall in a simple three tranche CDO

Thus we can imagine the situation as in figure 6.3. Losses come from the bottom up, eroding first the junior, then the mezz, then the senior tranches. Cash comes from the top down, paying first the senior, then the mezzanine, then the junior securities. Let's make the numbers a little more precise. Suppose the average credit spread of the bonds is 100 b.p.s: then if all of the bonds are current, we have $10M coming in per year. Assume that we have to pay the senior holder a generous 25 b.p.s, and 400 b.p.s goes to the mezzanine[185]. The CDO liabilities are then as shown in figure 6.4.

Thus even if we have ten defaults, and so there is only $9M a year in interest coming in, the junior will still receive $4.725M, corresponding to an interest rate of more than 11%. The *excess spread* of the collateral over that

| Tranche | Size ($M) | Spread (b.p.s) | Annual interest |
|---|---|---|---|
| Senior | 910 | 25 | $2.275M |
| Mezzanine | 50 | 400 | $2M |
| Junior | 40 | Residual | Interest paid by the bonds – Interest paid to other tranches |

**Figure 6.4**: Tranche details for a simplified investment grade CBO deal

needed to pay interest on the mezz and senior is therefore key to the value of the junior. If the excess spread is high enough, the junior can be a valuable tranche even if it does not repay any principal at all[186].

### 6.1.4 Prepayment

How do principal payments on the collateral affect the issued tranches? This is not a major issue for a CBO deal as the bonds are very unlikely to repay principal early. But if the collateral is prepayable, like mortgages or credit card receivables, then the question is important. There are two common choices:

1. In *sequential pay structures*, prepayments come from the top down. Thus the senior is paid first, then the mezz, finally the junior;

2. In *parallel pay structures*, each tranche is paid pro rata, so in our example tranching, a dollar of prepayments on the collateral would give ninety two cents of prepayment on the senior, five cents on the mezz, and four cents on the junior.

Some structures flip between these modes: they start off sequential pay, but then change to parallel pay.

### 6.1.5 Making it Better: Credit Enhancement

There are various ways of reducing the risk of any given tranche:

- The SPV can buy insurance on the performance of the collateral. For instance, it could find an insurance company which is willing to make good any losses between $40M and $90M, and thus remove the risk of the mezzanine tranche (assuming that the insurance is effective[187]). This form of credit enhancement is sometimes called a tranche *wrap*;

- Rather than giving all the surplus cash after the mezzanine and senior interest has been paid to the junior tranche, some of it can be kept in case of a high level of losses later. Thus a *spread account* is created which sits in the waterfall between the junior and mezzanine tranches, and is slowly filled as the collateral pays. This account is then available to meet later losses: if it is not necessary for this purpose, it is paid to the junior tranche holder at maturity;

- There may be *early amortisation provisions* where if bad things start to happen – for instance if loss levels are higher than expected – then rather than paying the junior, excess cash is used to prepay the senior tranche.

These techniques are known as *credit enhancement*: they are intended to improve the credit quality of one or more of the issued tranches. Notice that having sequential payment is a form of credit enhancement too as it gets the senior security off risk quickest: here not only does the junior absorb the first losses, it is around to take losses for longest.

In many cases the *attachment points* of the tranches – the amounts of losses they absorb – and the credit enhancement are arranged so that the senior tranche is as large as possible and is still AAA rated. Thus for instance the issuing bank in our CBO might discover that a $90M cushion was enough to get the senior to AAA. Similarly the junior is only as thick as it needs to be to get the desired rating for the mezz.

### 6.1.6   Why Securitise?

The obvious reason to securitise assets is to *transfer risk*. In the ODM, securitisation frees up capital and funding, allowing the firm to lend again. Providing that the securitisation is legally robust – the assets really have been sold to the securitisation SPV and the tranche holders have no claim against the originator – this is a perfectly reasonable strategy. As Northern Rock discovered to its cost, the ABS market is not always open, but certainly securitisation has allowed banks to provide more credit and cheaper credit than would otherwise be the case.

Investors may benefit here too in that securitisation allows them to invest in otherwise inaccessible assets. Without securitisation, if you want to take unsecured Japanese retail credit risk, you need a way of originating it. That requires a presence in Japan, advertising, a brand name, regulatory approvals and so on. But if a Japanese bank securitises its credit card portfolio, then a Western investor can get access to this class of investment without the costs associated with opening a Japanese credit card issuer.

Suppose investors really do believe that the senior tranche of a CBO is low risk. They are willing to receive a credit spread of ten basis points, so that if the senior ABS pays Libor + 10 b.p.s., then they will buy it at par (i.e. pay 100 for the right to receive 3 month Libor plus 0.1% for five years on 100, then get their 100 back). The originator sells all $910M of the senior tranche to investors, but buys both the mezzanine and junior tranches itself. In this situation the originating bank has kept the first $90M of losses, but passed on the risk of very large losses. There are two good reasons it might do this:

1.  The bank's own credit spread at five years is more than ten basis points. Here, then, the securitisation is basically a *funding arbitrage*: investors are willing to receive less credit spread for the risk of losses on the senior tranche than they require for buying the bank's unsecured five year bonds. The bank can fund the collateral, and keep most of the risk and return on it, more cheaply using securitisation than it could if it kept the collateral on its own balance sheet[188].

2.  Very nearly all of the risk of the securitisation in most circumstances is in the bottom tranches. Suppose regulators require capital equal to 8% of the notional held. Before securitisation therefore, our loans require capital of $80M. If a bank securitises as above but keeps the junior and the mezzanine tranches, it has a total notional of $90M, and therefore the capital requirement is 8% of that or $7.2M. Thus the capital has fallen by more than 90% but the risk has not changed very much, and neither has the return. The bank's return on capital increases dramatically: it has achieved a *regulatory capital arbitrage*. Regulators eventually closed this loophole, but it was important in drawing banks into the securitisation market in the late 1990s. More sophisticated regulatory capital arbitrages using securitisation still exist.

Notice that investors may well want originators to keep some of the lower tranches. The lowest tranche suffers losses first, so if this is retained by the originator, investors may believe that good quality assets will be put into the deal since losses hit the originator first[189].

If we can buy the collateral for less than we can receive for selling the tranches, then there is a *pure arbitrage*. The demand for diversified credit risk has in the past been sufficiently large that this was possible: one could start with nothing, borrow money, use it to buy bonds, securitise them, sell all the tranches, repay the loan, and be left with a substantial profit. So much for efficient markets.

Finally, another motive for securitisation is the profit available from ancillary transactions. For instance the underwriting bank will earn a fee from selling the tranched ABS. Profit is also available from doing transactions with the securitisation SPV[190]. Moreover if the collateral is in any way *managed*, asset management fees are available.

There are therefore a variety of different motivations for securitisation. Some of them are perfectly reasonable and make the market more efficient. Others, such as regulatory capital arbitrage, are less desirable. (One might argue, though, that the existence of these arbitrages is due to flaws in the regulatory regime. Banks have a duty to optimise their return on equity, and if supervisors allow them to do this via securitisation, then it is hardly surprising that they take advantage of this feature of the rules.)

### 6.1.7 The Rating of Securitisation Tranches

By 2006, securitisation had become an enormous business. Three trillion dollars of ABS were issued in that year, up from roughly a billion in 2002[191]. Structures had become complex, too, with more tranches, increasingly sophisticated forms of credit enhancement, and a widening range of collateral. How did investors understand what they were buying?

The sad answer appears to be that some of them didn't. Instead, they relied on credit ratings. Partly this was because of the Hunt for Yield described in section 3.4.9: faced with collapsing credit spreads, investors were forced towards more and more exotic instruments to earn a good return. But there was also a good measure of laziness. When they were offered a complicated instrument which appeared to offer a good return, some investors simply failed to try to understand what they were buying, taking comfort instead in a high credit rating.

Needless to say there were problems with this. One was that a single rating cannot possibly capture all of the behaviour of a complex financial instrument. As an example, consider two one year bonds, both paying a credit spread of 100 b.p.s:

- Bond 1 returns 100% of principal 99 times out of 100, but one percent of the time, investors lose everything;

- Bond 2 returns its full scheduled principal only 90% of the time, but in the remaining 10% of occasions, investors still receive 85% of their investment.

Which bond is riskier? On a pure expected loss basis, it is bond 2. But bond 1 – while mostly safe – can go very badly wrong for the investor. At least with bond 2, you never lose everything.

There is no right answer to the question of which bond is safer because there are many definitions of risk[192]. Depending on one's tolerance for unlikely but painful events, a reasonable case can be made for either bond 1 or bond 2 being riskier. A single rating cannot hope to capture the full richness of behaviour of ABS.

A second rather more serious problem is that the ratings agencies did not do as good a job as some investors thought they were doing[193]. Part of the reason for this was a lack of alignment of interests: the agencies are paid by issuers; but investors use the information. It is clearly in the interest of the party that pays to get as high a rating as possible. For corporate issuers, where ratings are relatively easy to determine independently, this lack of alignment of interests was offset by the reputational risk posed by over-generous ratings. But for ABS, the ratings process was more complex and reliant on difficult-to-source or non-public information: it was difficult for a third party to challenge the agencies.

The primary ABS market was dominated by a handful of banks and broker/dealers. These firms designed asset backed securities and underwrote their issue. The agencies had to keep these key clients happy as they derived more than twice as much revenue from rating ABS as from any other business line[194]. As part of this process, the agencies gave access to their models to the structuring banks. This meant that ABS structures could be optimised to get the best possible rating for each tranche.

Moreover, if one of the three major agencies had a particularly generous treatment of a given feature, they would naturally be used for rating securities with that feature, a process known as *ratings shopping*. The structuring banks were able to exploit any flaws in the rating process, and the almost complete absence of unsolicited ABS ratings (where an Agency rates a security that it has not been asked – and paid – to rate) meant that investors almost always saw the best rating available.

## 6.2   The Securitisation of Subprime Mortgages

Securitisation was by far the most important way of funding subprime lending: over three quarters of subprime loans were securitised in 2005 and in 2006[195]. The basic techniques of transfer to an SPV and the use of a waterfall to create tranched securities are the same as other securitisations. However there are some structural features of subprime-backed ABS that are unique, and which make the value of the tranches very sensitive to house prices[196].

### 6.2.1  The Structure of Subprime Securitisations

Subprime deals typically have a complex tranching structure. The two most common arrangements during the Boom years were:

1. *Six pack* deals, where the mezz is divided into six tranches (for instance rated AA, A+, A-, BBB, BB, and B); or

2. Three or four tranche mezz deals with an excess spread account.

In both cases, the transaction usually starts out as sequential pay, then flips to parallel pay either after a fixed period, or if the deal performs well[197].

### 6.2.2  Sweet and Far from Cliff and Scar

If the underlying mortgages perform, then a tranche of a subprime securitisation becomes better rather quickly. Most of the mortgages are 2/28 or 3/27 teaser ARMs, so there is a huge wave of prepayments in years two and three. These prepay the senior, so the top tranche might have a life of only a couple of years, and hence is not exposed to default risk for long. In deals with an excess spread account, the relatively high teaser rate and prepayment penalties of the underlying mortgages mean that the spread account is funded fairly quickly, increasing the quality of the mezz tranches as time goes on.

The problem is that these good things are racing against bad things: delinquencies and defaults. The more delinquencies there are, the less spread there is. The more defaults there are, the lower the level of prepayments. Moreover in a subprime deal what defaults there will be tend to happen early, due to the need to refinance the loan after the teaser ends. At this point a wave of underperforming loans can overcome the credit enhancement in the structure and send even the price of the senior securities off a cliff[198].

### 6.2.3  The ABX

We first came across the ABX indices in chapter 1: now it is clear what they are. Each index is comprised of tranched subprime ABS, so the ABX 0602 BBB is an index of BBB-rated subprime-backed bonds issued in the second half of 2006. The indices provided visibility of the market price of subprime risk as processed by the securitisation structures. House prices fell, defaults rose, and the credit enhancement in subprime deals became less effective. The senior securities relied on this credit enhancement for their performance, so their prices – along with all the other tranches – fell fast. Everyone could see this as the ABX prices were widely distributed.

# 6.3   Models and Hedging

A *derivative* is a financial instrument whose value depends on something else, most commonly another financial instrument. Thus for instance if I have the right to buy ten shares of a bank at $10 per share, the value of that right depends on the bank's share price. If the share price is over $10, the derivative is clearly valuable.

Many derivatives are traded infrequently. Once a dealer has sold such a thing, they must assume that they cannot buy it back. This gives rise to risk: suppose a dealer has sold me the right to buy ten shares for $10, and I can exercise that right any time up to December 2009. If the price of the *underlying* shares go up, my option is worth more, and so the dealer loses money. If the underlying goes down, they make money.

Most derivatives dealers wish to constrain the size of possible losses due to market movements. They cannot buy back their exact exposure. Therefore they *hedge*. Hedging is the process of trying to immunise a portfolio against possible losses caused by market moves: typically it involves either giving up profits in exchange for loss protection, paying a fee, or both.

Hedging is central to capital markets trading for two reasons:

- Most banks try to hedge their portfolios, at least partially; and
- The valuation of derivatives is based on the hedging process.

## 6.3.1   The Idea of Hedging

Let us examine our example derivative in more detail. Suppose it is Wednesday 1st July 2009, and I have just bought the right to buy ten shares of DoublePlusBank for $10: this is known as a *call option* on DoublePlusBank. I can exercise my right at any time between now and 31st December 2009, simply by calling the option seller. I tell them that I wish to exercise, pay one hundred dollars, and receive my ten shares.

Obviously the more the underlying ten shares are worth, the more the option is worth: I am not going to pay one hundred dollars for something that is worth eighty, so I will not exercise the option if the share price is eight dollars.

Figure 6.5 shows the variation of the price of my call option today as the share price of DoublePlusBank changes. The more it goes up, the more my option is worth. Notice too that even if the DoublePlusBank's share price is below ten dollars, the option is worth something. This is because even though it is not worth exercising the option now, it might be in the future: we have six months for the share price to rise above ten dollars. Suppose

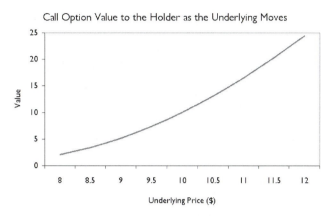

**Figure 6.5**: The variation of the value of a call option as the underlying varies

DoublePlusBank's shares now are trading at $9. Then the option is then worth roughly $5: this is what I would pay the dealer for it.

The dealers' profit and loss (or P/L) as the underlying share price moves is as shown in figure 6.6: if the option is worth more than $5, the dealer loses money; whereas if it is worth less than $5, they have made money.

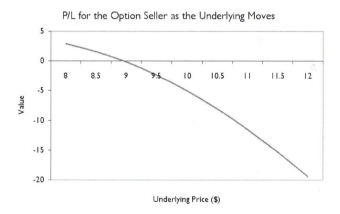

**Figure 6.6**: The P/L for the option seller as the underlying varies

Clearly in order to avoid the situation of losing far more than the value of the option, the dealer needs to hedge. If the underlying goes up, they lose money. Therefore a hedge will be something that makes money as the underlying goes up. One thing that makes money as the underlying goes up is a position in the underlying itself. Therefore the dealer can hedge by buying shares in DoublePlusBank. Specifically if the underlying goes up from $9 to $10, the dealer loses $5.02. Therefore the dealer needs enough shares to make $5.02: five shares would make $5 for such a move, so a reasonable hedge is to buy five shares. The dealer therefore purchases five shares in DoublePlusBank. The resulting hedged portfolio – the option they have sold plus the shares – is much less sensitive to movements in the share price as figure 6.7 illustrates[199].

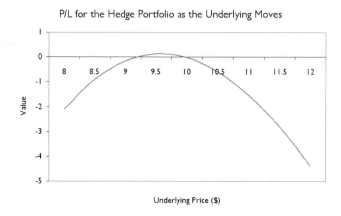

**Figure 6.7**: The value of the hedged portfolio as the underlying varies

Successful hedging therefore ensures that dealers do not lose (or make) too much money as the market moves. But how can we decide if we are hedged?

### 6.3.2   Models in Finance

There are three types of model typically used in finance:

1.   Pricing models;

2.   Hedging models; and

3.   Risk models.

*Pricing models* are the simplest. Here we know the prices of certain liquid benchmark instruments, and we simply want to *interpolate* between those prices. Thus for instance certain call options on DoublePlusBank expiring on 31st December 2009 might trade on a stock exchange: perhaps the exchange quotes the price of an option to buy the stock for $9 per share, and that of another to buy for $11.

We want to price an option to buy the stock at $10 per share. A pricing model does this. It must reproduce the prices of the benchmarks, and it should also give us prices away from the benchmarks which reflect the level at which we can genuinely trade[200]. We are in no sense predicting the *fundamental* value here: we are simply assuming that the given benchmark prices are correct, and deriving a price for a related instrument in that context.

In contrast, many instruments will only be traded a handful of times during their life: some trade only once. Here, then, the price at which we could trade the instrument is in a sense irrelevant, as we *won't* actually trade it. Rather what we care about is the amount of money that can be locked in through hedging. In this situation, a dealer wishes to sell something for more than it costs to hedge over its life, whatever happens to the markets (or to buy it for less than it will cost to hedge).

Standard derivatives models calculate exactly this. That is, under certain assumptions, they estimate *the cost of hedging* a derivative. The 'fair value' of such instruments is identified with the (model's prediction of the) cost of hedging it[201]. Therefore what we want from a *hedging model* is for it to generate hedge ratios – amounts of the underlying to buy or sell – which reduce risk as much as possible. If such a model is successful, and we hedge diligently, then any profit we make from the original purchase or sale will not be eliminated by subsequent market movements.

The same model is often used both for pricing liquid products (short dated options for instance) and for hedging longer dated products. For the former, the price the firm uses for the product is regularly validated by actual market transactions. For the latter, the key issue is whether the price the model suggests is one that the firm can actually realise by hedging according to the model hedge ratios.

| Final Stock Price | Probability of Moving at Least that Far | Loss |
|---:|:---:|:---:|
| $10 | 25% | $5 |
| $11 | 8.5% | $11.5 |
| $12 | 2% | $20 |

**Figure 6.8:** Probabilities of DoublePlusBank stock moves up and the resulting loss for the unhedged option position

Finally, *risk models* attempt to estimate how much a portfolio can lose in given situations. Typically such models assume that the portfolio is correctly valued to start with, then attempt to estimate how likely various losses are. For instance for the unhedged option position, if we estimate that the probabilities of moves to $10, $11 and $12 are as shown in figure 6.8, then we can for instance infer that there is a one in four chance of losing five dollars or more, and a one in fifty chance of losing $20 or more.

Risk models typically have market volatilities as inputs: the more the markets move, the more likely large losses are. Another input is correlation: the extent to which markets move together. If correlations are high, markets tend to move together, and losses can be larger than when market moves diversify each other.

### 6.3.3   How Financial Models Differ

The term 'model' can imply something sophisticated: one may think of science, where a model might be based on an accurate theory of the world and hence have predictive power. The theory of gravity gives a model of what happens when we drop a cannonball from a tower, and this model predicts how long it takes the cannonball to fall and how fast it is going when it hits the ground. Pricing, hedging and risk models predict nothing: they simply tell us respectively how much something is worth given the benchmark prices; how much we should expect it to cost to hedge; and how likely a given loss is.

If that were all, things would not be too bad. But there are two further vital differences between scientific and financial models:

1.  The model *can change the behaviour*. In science, a better model does not change the way the world works. The same experiment gives the same result regardless of the theory being tested. But once a model is accepted in finance, it is used to make trading decisions. Those trading decisions affect the dynamics of the market. Thus the theory actually alters the behaviour[202].

2.  The model is *calibrated to current data*. The constants in scientific models are genuinely constant: the weight of our cannonball might be an input to the model, but its weight is readily determinable and invariant (until we drop it and it breaks, anyway). Many financial models in contrast have an input which is based on market behaviour. For instance volatility – how much prices move around – is a key input to derivatives pricing models. But volatility is not constant. It takes different values depending on when we look and how long we look for.

The usefulness of financial models can therefore be limited to a particular period in the markets. During that period – when the market is working as they conjecture it does – they are fine. Calibration is easy and the success of the model may well encourage its wide acceptance. But then a sudden change in market conditions can mean that the model works much less well. For hedging models, this can mean that the model's purported hedge does not in fact work, and there is a large loss as a result. (A large profit is also possible, but this happens rather less often.) For risk models, it means that actual losses can be much bigger than predicted ones. Storms that the model predicts should occur no more than once in a thousand years in fact hit several times in a decade[203].

### 6.3.4   ABS Models and Correlated Defaults

One key issue for models of ABS is how likely large numbers of defaults on the collateral are. Consider the 100 bond CBO deal discussed earlier in the chapter. If we make the assumption that the credit spread of each bond is compensation for the probability of default, then we can figure out how likely the market thinks each bond issuer is to default based on that credit spread. But bond issuers are not independent: for instance, if Ford defaults, it may be because of bad economic conditions, and these will also affect GM. Ford's default is then said to be *correlated* with that of GM.

The more correlated bond issuers are, the more dangerous the senior tranche of a CBO is. One way to think about this is to imagine a minefield. Each bond corresponds to a mine, and a default is the mine blowing up. If defaults are highly correlated, then the mines are clustered close together. This means that if we have already seen one go up, we are likely to be very close to others, and the chance of another one or two explosions is large. If they are less correlated, the mines are further apart, and a single explosion does not significantly increase the chance of further bangs.

It takes quite a few defaults to breech the senior tranche, so its value is heavily dependent on the chances of multiple defaults. The more correlated defaults are, the less safe the senior tranche is[204].

Figure 6.9 illustrates this: for the same bonds, as defaults become more correlated, the probability of a large number of defaults in the 100 bond CBO increases dramatically.

## 6.4   Model Risk

Financial instruments whose value is derived from a model form a significant fraction of the balance sheet of some firms. This is especially true in a crisis: here liquidity disappears and market quotes are not available. Determining fair values then involves *mark to model*. The mistakes that can arise here are known as *Model Risk*[205].

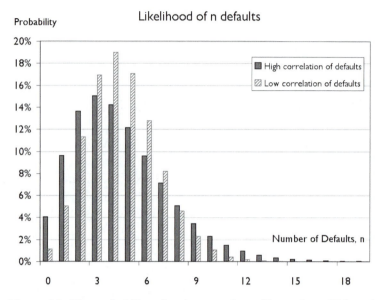

**Figure 6.9:** The probability of a given number of losses in a CBO as defaults become more correlated

To see these forces in action, consider a large broker dealer. In the second quarter of 2008, Goldman Sachs had a total of $640B of assets valued using a model, out of a gross total of $765B (and supported by shareholders' funds of $43B)[206]. The situation is similar for other large capital markets players: JPMorgan, for instance, used models of 86% of its trading assets in a similar period[207]. Where, then, might we wish to be careful in the use of models?

### 6.4.1   Elementary Mistakes

Models are encoded in computer systems, and those systems feed financial reporting. Therefore all of the mistakes that can be made in the use of computer systems can be made in financial models. These include:

- Coding errors. The model is mis-coded and does not do what is intended;

- Data ambiguity errors. The model expects one kind of input, but is actually fed another. Common mistakes here include dates (American style MM/DD/YY vs. European style DD/MM/YY) and relative vs. absolute moves (does a 10% move up from 30% take us to 40% or to 33%?);

- Arithmetic or convergence errors. Models often use numerical simulations of complex mathematics. This process can go wrong in subtle ways relating to the precise implementation of the numerical procedure[208].

Typically – although not always – these problems are caught by a *model review process* which ensures that before models are used for the firm's financial reporting they are properly validated.

### 6.4.2 Calibration

Most financial models are calibrated somehow or other. That process might be very simple: for instance it may just be necessary to enter a Libor rate. Or it might be rather complex, involving hundreds of pieces of market data. In any case the model is no better than its calibration. The issue here is not so much that the inputs might be wrong but rather that they might be uncertain. If a bond has not traded recently, it may be difficult to be sure of where it would trade today should we be lucky enough to observe a transaction[209]. Best practice here is to take valuation adjustments which reflect the magnitude of the uncertainty. This allows the institution to use their best guess of the value of an instrument but keep a sum aside in case that estimate is in error. If we do not know whether a credit spread input should be 100 b.p.s. or 120, and there is a \$2M difference in value depending on which number we pick, then we should use 110 b.p.s., and take a \$1M valuation adjustment. The size of this valuation adjustment is a reminder of the inherent uncertainty in the fair value of the portfolio concerned.

### 6.4.3 Theory and Induction

The next class of model risk issues are more dangerous because they are harder to mitigate. First, there are choices in the mathematical model used to describe many more complex products. These mathematical choices give rise to different models, with different prices and different hedge ratios for the same instrument[210]. There is often no particularly good reason to prefer one of these choices over another: all of the models considered are sophisticated; they are all based on a plausible idea of how the markets move, and they all calibrate correctly to the benchmark instruments. They just happen to differ as to what certain derivatives are worth.

The other real problem is an implicit assumption that the future is like the past. When we model a cannonball being dropped, the assumption that gravity works the same way today as it did yesterday is a reasonable one. In finance, the assumption that the FTSE-100 index will move the same way today as it did yesterday is much less easy to justify, not least because there is good evidence that over time its dynamics do change significantly[211]. In particular in a crisis, assumptions that were reasonable for many years are suddenly false. The usual relationships break down[212], and so models based on the truth of these relationships can fail.

Risk models are susceptible to this form of breakdown too. They are calibrated using data from recent periods. If the future is very different from the past – as it is when a crisis hits – their predictions can be very unreliable. In other words, risk models are least reliable when firms need them most[213].

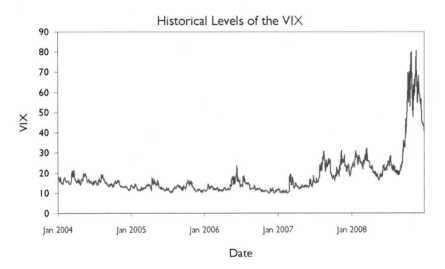

**Figure 6.10**: Market expectations of S&P 500 volatility as a function of time

Figure 6.10 illustrates this: it shows the market's expectations of short term volatility for one of the main U.S. equity indices, the S&P 500. Clearly if we have a model that assumes volatility is constant – as many risk models do – and the model is calibrated to the low level of volatility which pertained in the 2004-2006 period, then we would grossly underestimate the probability of large moves: large moves which were actually seen in 2007 and 2008. Thus using a risk model *with a calibration to recent history* is tantamount to assuming the future will be like the past. Sometimes, it isn't.

### 6.4.4   Stress Testing

It is well known that financial models do not work in all conditions. Therefore firms are required to *stress test* their portfolios, that is to subject them to an imagined stress and review the results. If this process is done diligently and with considerable imagination, it can highlight model assumptions and the consequences of their failure. But it is easy to stress test for the last crisis – against which the portfolio will inevitably be well hedged – and to miss the specific features which make the next one toxic. Given that no one can seriously be expected to foresee everything that might possibly go wrong and test for it, stress tests are inevitably incomplete in their coverage[214].

### 6.4.5 A Category Error

One common mistake is to think that a model price or a model risk estimate is somehow an objective thing independent of market conditions and unburdened with assumptions. Model outputs may be seen as statements about reality. Even when the models concerned are honestly developed and properly verified, the veracity of model inferences depends heavily on normal market conditions. To use a description from an analysis of model risk in another setting[215]:

> The numbers [that come out of a model] have no meaning
> except as illustrations of the consequences that flow from the
> model's assumptions.

The ultimate defence against model risk, then, may simply be a better understanding of what a model is capable of telling you.

Just before the Great Depression, Frank Knight made the distinction between 'risk' and 'uncertainty'[216]. For him, risk was when possible futures are quantifiable, so there is a risk when calling heads that the coin will come down tails: it happens 50% of the time with a fair coin. Uncertainty, on the other hand, happens when we do not know how likely a future state is and cannot meaningfully estimate that likelihood.

The problem in finance is that a lot of the time there is actually uncertainty, but we convince ourselves there is risk. When model risk bites, and we find that our estimates of future probabilities are wrong[217], the true uncertainty of the situation becomes apparent. This typically causes a loss of confidence, a general reduction in risk taking, and a flight to quality. Even perfectly good bets become suspect in this context, simply because market participants doubt their own ability to assess what a good bet is. Of course, they never could. But false confidence encouraged excessive risk taking, and the realisation of the true situation causes something close to panic.

## 6.5   Where Did It All Go Wrong?

The transmission of small or medium-sized risks around the financial system is clearly a good thing: no one institution is endangered by the risk, and confidence is maintained. It may be, though, that a very large risk is best not distributed. This is because if everyone mis-assesses a risk, everyone (or at least many firms) will take that risk. The more of the risk is distributed, the more the system suffers. Without distribution, individual institutions might fail, but these would be the originators: others could not access this mis-priced risk.

This happened in the Crunch: the risks of mortgage lending were misjudged. This was partly because historic data indicated that mortgages were safe. The historical data did not include the effects of increasingly risky lending, and nor did it give insight into the effect of a fall in house prices. ABS then transmitted this risk around the financial system[218]. Various factors then conspired to make the situation worse.

### 6.5.1    Intensifiers: Complexity and Leverage

The more complicated something is, the more we rely on models. A single loan can be assessed by a good loan officer. The risk of a pool of thousands of loans can, with diligent work and if reliable information is available, also be assessed. But by the time we get to the risk of a tranche of a securitisation with two different kinds of credit enhancement, a model is necessary to assess risk. With models come model risk, and typically the more complex a security is, the more risk there is assessing it.

Once something is packaged as a security, leverage becomes easy. We can buy it and finance it on a repo basis for instance, and hence we only have to find the repo haircut rather than the full lump sum. Leverage again intensifies risk: a 5% mis-estimate of value on funded risk means a 5% loss; but if the position is repo'd with a 5% haircut, then the same error means everything is lost.

### 6.5.2    Underwriting Standards

Securitisation enables sloppiness. Specifically, if a firm knows that the risk of any loans they make will end up with someone else, there is little incentive to lend to people that are likely to repay the loan. This is especially the case if their return depends on the volume rather than the quality of the loans originated.

There is a simple way to fix this problem: ensure that originators keep a minimum amount of the loans they make. The EU even tried to do this: they proposed that banks should keep at least 15%, itself a modest amount. Comment from ODM banks was vociferous, and the minimum was watered down to 10%[219].

Thus the incentive structure set up by securitisation; the leverage that was available to buyers of ABS; and the complexity of the securities, all meant that subprime ABS positions were highly risky. These securities were also, as we saw earlier, peculiarly exposed to falling house prices. Why did so many professionals fail to see the problem?

### 6.5.3   Model Error in the Assessment of Subprime Securitisations

The main error that the ratings agencies and many others made in assessing the quality of subprime securitisations was only to contemplate changes in market conditions which were consistent with the recent past. Thus in order to get a AAA rating, a security had to be robust under delinquency rates which were small multiples of those observed in 2000-2005. Delinquency rates ten or twenty times higher were not contemplated. This is because most risk assessment models had been calibrated to a period of rising house prices: modellers literally had no idea what would happen if house prices started to fall. This hidden risk factor which would dramatically increase default rates was not considered[220].

The type of credit enhancement in subprime ABS made this error worse: excess spread and tranching only protect the more senior securities if defaults do not happen too soon (so that excess spread is available) and there are not too many of them. Thus the senior tranches of subprime ABS transactions were only AAA when house prices were rising. When they started to fall, a wave of defaults hit the structure, and the credit enhancement available was not judged to be sufficient to save the senior tranche[221]. Investors suddenly realised that the models used to assess subprime ABS were deficient, and rushed to sell the securities. Both the price and the credit quality of subprime ABS fell, with only 25% of tranches issued in 2005-2007 and originally rated AAA retaining that rating by mid 2008[222].

A final issue in the rating process was the choice of risk indicators. Historically high FICO borrowers were much less likely to default than low FICO ones, and the relationship between FICO and default rate was reasonably stable. FICO was therefore an important input into the ratings process. But this relationship broke down in the current crisis: recently LTV and PTI – that is, how much equity the mortgage holder has in the home and how affordable the mortgage is – have been a better indicator of default rates than FICO.

### 6.5.4   A Waxman Pecora?

The House of Representatives Committee on Oversight and Government Reform has been conducting a series of hearings into various key events during the Crunch that echo the Pecora hearings. The Committee's ability to persuade individuals to testify and to obtain the release of confidential or hard-to-discover documents has revealed much about the state of American financial services. We will refer to it as the Waxman Committee after its chairman, Representative Henry Waxman.

Various pieces of testimony given to the Waxman Committee will be discussed as we discuss the relevant events: for now, Edward Pinto's testimony is relevant. He said that[223]:

> In 1992 a mortgage borrower with a FICO of 620-659 was seven times more likely to experience a serious delinquency over the next two years than a borrower with a 720-759 FICO. By about 2004, the 620-659 borrower was now twelve times more likely [to experience a serious delinquency] and the default propensity of the 720-759 borrower was unchanged.

In retrospect, then, the subprime risk assessment process did not correctly weight even the little data that was available.

## 6.6   The Write-downs

By the end of September 2008, the subprime-related losses of the world's 100 largest banks had reached nearly $600B[224]. The largest losses came from some of the largest banks: figure 6.11 shows the five firms that lost over $40B each.

| Firm | Writedown & Loss |
|---|---|
| Citigroup | $60.8B |
| Wachovia | $52.7B |
| Merrill Lynch | $52.2B |
| Washington Mutual | $45.6B |
| UBS | $44.2B |

Figure 6.11: Estimates of the largest five subprime-related losses in banks as at the end of September 2008

Two of these names are not surprising: Wachovia and Washington Mutual were both highly active in the American mortgage market, so their exposure comes as a natural part of their business. But how did UBS – a bank that did not even offer subprime mortgages – and Merrill Lynch – a broker/dealer – end up with such big losses?

### 6.6.1 Funding Arbitrage

Consider a very large bank. Before the Crunch, the best and biggest funded slightly *under* Libor. This put them in a unique position compared to even slightly smaller banks: they could buy a security paying only a few basis points more than Libor and make money. Thus for instance a bank which could fund at Libor minus five could buy a bond paying Libor plus five and make ten basis points from it. A bank which funded at Libor plus eight would suffer a three basis point loss per year on the same position.

The AAA-rated tranches of securitisations paid only a small spread to Libor: perhaps high single digit or teens of basis points. Thus they were only sensible investments for firms which funded close to or beneath Libor. These players had a funding arbitrage: they could raise unsecured debt and use the proceeds to buy AAA-rated ABS. Thus the senior tranches often ended up being bought by the banks with the lowest cost of funds: that explains why Citigroup, Merrill Lynch and UBS had such large positions. Each of these firms could sometimes fund below Libor. Model risk affected their assessment of ABS. Thinking that they were very low risk, they bought billions of dollars of AAA tranches, exploiting their low cost of funds to make what may have appeared to be a risk free spread. This spread was not large – perhaps only ten basis points – but 0.1% on $50B is $50M a year[225].

It is worth noting here that the AAA tranche is a large percentage of the securitisation: often over three quarters of the total. The roughly $400B of securitised subprime mortgages therefore generated something in the neighbourhood of $300B of AAA tranches[226]. Much of this risk was funded by the largest banks[227]: they were more or less the only people who could hold this risk profitably.

### 6.6.2 ABS Liquidity Risk

When the Crunch hit, the prices of tranched ABS fell fast. Liquidity also disappeared. Lacking confidence in ratings, and without the information needed to do their own analysis, bond buyers ignored ABS. This meant that the banks could not sell their inventory positions. Those that had large positions were forced to take write-down after write-down: selling or hedging a multi-billion dollar position in a liquidity crunch is very difficult.

As an example of the sequence of losses, see figure 6.12. This shows Citigroup's write-downs for the last quarter of 2007 and the first two quarters of 2008 on subprime[228].

If these numbers look large, that is because they are, both in the abstract and compared with the profits the firms made. One estimate for Merrill Lynch

| Quarter | Write-down |
|---------|-----------|
| 4Q 2008 | $5.6B |
| 3Q 2008 | $1.7B |
| 2Q 2008 | $5.9B |
| 1Q 2008 | $7.5B |
| 4Q 2007 | $18.1B |

**Figure 6.12:** Citigroup total write-downs on sub-prime related direct exposures

is that the sub-prime related write-downs that the firm suffered in 2007 and 2008 eliminated a quarter of all the profit the firm made in its thirty six years as a listed company[229].

### 6.6.3   How Wrong were the Firms' Risk Estimates?

Let us take Citigroup's losses, and turn them into a loss per day: a $60.8B total loss works out at roughly $200M per day for each day since the write-downs began. Citigroup's risk estimate of the amount it was likely to lose no more than one day in a hundred was $142M. On average, then, the bank lost more than its risk estimate every single day from the start of the Crunch to late 2008[230]. This illustrates how inaccurate risk models can be in a crisis.

### 6.6.4   The Consequences of the Write-downs

ABS trading is a game of pass the parcel conducted in the dark. When the music stops, you do not know who has the parcel. In the Crunch it became clear that holding the parcel was a really bad thing: the write-downs in late 2007 were evidence enough of that.

But were there players who had subprime risk but who had not written it down? The markets were concerned that there were[231]. At this point, the opacity of the disclosure of financial institutions became a handicap rather than an advantage. Firms scrambled to convince investors that they did not have subprime exposure or, if they did, that it was contained and properly valued. Investors were sceptical, and the equity prices of any firm which could not prove that it was squeaky clean fell.

Ratings agencies, seeing the losses, downgraded many of the firms involved. Repeated assurances that the worst was over followed by further losses sapped confidence. The stage was set for the next phase of the Crunch.

# Notes

[181] The phrase is attributed to Richard Fisher, President of the Federal Reserve Bank of Dallas.

[182] Unless stated otherwise, by MBS we mean 'residential mortgage backed security', something some people call 'RMBS' in distinction to 'CMBS' for a security backed by commercial property loans.

[183] This distribution also depends on assumptions about how defaults are associated: that is, how likely they are to happen together. Here a simple model of default association was used: we discuss this issue in section 6.3.

[184] The account here is simplified. First, we have ignored some cashflows that get paid first, such as fees for the note trustee. Second, we have assumed one waterfall for principal and interest, whereas in reality there may be separate P and I waterfalls. Third, there will probably be extra features in the structure, such as liquidity facilities or interest rate derivatives, and these may have claims at various positions within the waterfalls.

[185] This is a simplified structure: in reality there would be four or more tranches plus various other structural features. The basic economic argument would however still hold.

[186] This is particularly noticeable in credit card securitisations: there the excess spread is so large that part of it can be securitised as a separate tranche known as a *credit enhancing interest only strip*. See *Characteristics of Securitizations that Determine Issuers' Retention of the Risks of the Securitized Assets* by Weitzu Chen et al. (2007), available at `papers.ssrn.com/sol3/ papers.cfm?abstract_id=1077798`.

[187] Just as home owners sometimes find their household insurance coverage is rather less effective than they had thought when they actually come to claim on it, so too capital markets investors have sometimes found that insurance written on financial contracts does not behave the way they expect. See for instance *Use of Insurance Policies as Credit Enhancement in Structured Finance*, Fitch Ratings 2001.

[188] The problem with this example is that if the bank actually did this as outlined, it would not be able to deconsolidate the issuing SPV. Depending on the precise accounting regime, there are ways to solve this problem, so accounting issues do not always prevent securitisation being used to reduce a bank's cost of funds, but more complex structuring is required. We discuss this a little more in chapter 10.

[189] This belief is often misguided: the equity tranche receives the excess carry from all the structure whether or not it returns any principal. Thus beyond a given point, the equity investor actually wants the collateral to be as risky as possible, since that produces the biggest excess spread. This is particularly an issue in managed structures.

[190] Such as interest rate or cross currency swaps or providing liquidity facilities.

[191] Data from Box 2.1 in the IMF's *Global Financial Stability Report*, April 2008, available at `www.imf.org/external/pubs/ft/gfsr/2008/01/index.htm`.

[192] An economist would say that there are many possible *risk aversions*. See Matthew Rabin and Richard Thaler's Anomalies column *Risk Aversion* in the Journal of Economic Perspectives (Volume 15, Number 1 Winter 2001) available at `www.behaviouralfinance.net/risk/ RaTh01.pdf` and the references therein for more details.

[193] For a comprehensive account, see *Where Did the Risk Go? How Misapplied Bond Ratings Cause Mortgage Backed Securities and Collateralized Debt Obligation Market Disruptions*

by Joseph Mason and Josh Rosner available at `papers.ssrn.com/sol3/papers.cfm?abstract_id=1027475` or, for pre-Crunch analysis, *Enhancing the Accountability of the Credit Rating Agencies* by Stéphane Rousseau available at `www.rotman.utoronto.ca/cmi/papers/CRA_Study_Rousseau.pdf`.

[194] See Mason and Rosner for more details.

[195] 81.2% in 2005 and 80.5% in 2006 according to *The 2007 Mortgage Market Statistical Manual*.

[196] The discussion here is based on Gary Gorton's account in *The Panic of 2007+*, a paper prepared for the 2008 Jackson Hole Conference and available at `www.kc.frb.org/publicat/sympos/2008/Gorton.08.04.08.pdf`, to which the reader is referred for a much more comprehensive account.

[197] See Adam Ashcraft and Til Schuermann, *Understanding the Securitization of Subprime Mortgage Credit* (Federal Reserve Bank of New York Staff Report No. 318, 2008) for more details.

[198] Prepayment penalties often apply during the teaser period for hybrid ARMs, so the stress on the mortgage is entirely concentrated at the refinance date. Before that it is too expensive for most borrowers to prepay: after it the mortgage may well be unaffordable.

[199] The hedging process we have described is known as *delta* hedging. For more details, see chapter 2 of my book *Understanding Risk*.

[200] Thus an interest rate swap model which builds the Libor curve in a standard way is an example of a pricing model: certain benchmark swaps are liquid, and we are interested in the price at which we could trade away from those benchmarks.

[201] Observers with different risk aversions differ on the right price for a risky position: the only thing they can agree on is the right price for a risk free position. Therefore the only thing a derivatives pricing model can do is to derive the price of constructing such an instantaneously risk free position. Thus for instance the standard Black Scholes formula for pricing options on a single risky underlying does that by considering the cost of maintaining such a risk free position.

[202] This is the reflexivity of section 3.4.6 again.

[203] A good example of this phenomenon occurred to Morgan Stanley in the quarter ending 31st August 2007: their risk estimate at a 99% confidence interval was roughly $125M (i.e. would expect to lose more than $125M no more often than one day in a hundred), yet one day in the quarter they had a trading loss of $390M according to their 10-Q. Clearly the statement 'on average we will lose more than $125M no more than one day in a hundred' does not preclude losing $390M. Nevertheless, the actual loss is so much bigger than the probabilistic estimate that it suggests additional risk information is needed to understand the potential losses in the tail of Morgan Stanley's earnings distribution.

[204] The term 'default correlation' is a standard one in credit risk modelling. It is used as a modelling convenience: we assume some mathematical form to model the defaults of multiple issues – technically a *copula* – and then calibrate this model to the observed market prices of tranched ABS. For more details, see my book *Understanding Risk*, or Lee McGinty et al.'s *Credit Correlation: A Guide* ( JPMorgan Research 2004). 'Default correlation' is therefore an epiphenomenon: by saying 'the default correlation in this CBO is 0.3' what we mean is 'if we put 0.3 into a model of the value of a tranches of this CBO, we get the right price'. To suggest that there is a literal, real world correlation between the defaults is clearly meaningless: we could never test such a statement (or at least we could not test it without reference to a model of how default happens). Defaults happen only once, after all.

[205] For a further discussion of model risk in a financial context, see Emanuel Derman's *Model Risk* (Goldman Sachs Quantitative Strategies Research Notes 1996) available at www.ederman.com/new/docs/gs-model_risk.pdf.

[206] Netting and collateral reduce the gross total to a net of $676B. Both Level 2 and level 3 assets have been counted as 'valued using a model': see chapter 10 for more details of these levels. Data is taken from the Goldman Sachs 10-Q for the quarter ending on the 30th May 2008, available at www2.goldmansachs.com/our-firm/investors/financials/current/10q/10-q-2-quarter-2008.pdf.

[207] Specifically JPMorgan used Level 2 or level 3 for 86% of its assets accounted for at fair value according to its 10-Q for the quarter ending on the 30th June 2008. Since JPMorgan has substantial commercial and retail banking activities as well as capital markets undertakings, non-fair value assets are a much larger percentage of its total assets than for Goldman Sachs.

[208] For a mathematical account of some of the typical kinds of problem, see Chapter 5 of Morton and Mayers' *Numerical Solution of Partial Differential Equations* (2005).

[209] For derivatives hedging models, the issues are even more delicate. Here we will typically need an implied volatility as input, and that may or may not be easily observable. If the model requires many inputs – such as a local volatility model – then calibration becomes more of an art than a science. For a further discussion, see *Managing model risk* by Riccardo Rebonato, in *Mastering Risk* (2001).

[210] For an insightful account of the issues around different model choices, see *A Perfect Calibration ! Now What ?* by Wim Schoutens et al. (2003), available at ucs.kuleuven.be/research/reports/2003/report2003_03.pdf.

[211] For instance Edgar Peters makes the point in *Chaos and Order in the Capital Markets* (1996) that even the question 'Is it reasonable to attribute a volatility to major equity indices?' can often be answered 'no' at a good level of confidence.

[212] Arguments based on a lack of arbitrage require not just the theoretical ability to trade the arbitrage but the continuing presence of arbitrageurs. In a crisis, arbitrageurs in many markets are scarce, so these arguments can and do break down.

[213] For a longer account of this phenomenon see *The Emperor has no Clothes: Limits to Risk Modelling* by Jón Daníelsson, Journal of Banking and Finance (Volume 26, Number 7, 2002), available via risk.lse.ac.uk/rr/files/JD-00-6-10-960591721-1.pdf.

[214] The President's Working Group on Financial Markets *Policy Statement on Financial Market Developments* (March 2008) available at www.treasury.gov/press/releases/reports/pwgpolicystatemktturmoil_03122008.pdf mentions 'the failure of stress testing procedures to identify institutions' vulnerabilities to system-wide shocks to markets and market participants' while Morgan Stanley's CFO, Colm Kelleher, in a Bloomberg interview on 28th January 2008 said: 'Our assumptions included what at the time was deemed to be a worst-case scenario. History has proven that the worst-case scenario [we considered] was not the worst case'.

[215] This quotation comes from *Six (or so) things you can do with a bad model* by James Hodges (Operations Research, Volume 39, Number 3, 1991) available at www.rand.org/pubs/notes/2005/N3381.pdf.

[216] Frank Knight, *Risk, Uncertainty and Profit* (1921).

[217] The occurrence of 'once in a million years' events every decade is a bit of a giveaway.

[218] See *Credit Risk Transfer: Developments from 2005 to 2007*, BIS (July 2008), available at www.bis.org/publ/joint18.pdf for a much more comprehensive discussion.

[219]See the European Commission's *Second public consultation on possible changes to the Capital Requirements Directive* available via ec.europa.eu/internal_market/bank/regcapit al/index_en.htm.

[220]As Mason and Rosner point out, pooling of risk only creates diversification if there is no systematic risk factor which can affect most of the pool at once. With subprime, there was.

[221]Defaults have not climbed sufficiently to hit the AAA yet in many transactions. For instance, the first write-down on the ABX BBB- 0602 did not happen until June 26th 2008. The AAAs fell so far in value in 2007 because the market anticipated lots of defaults in the future, not because they were actually happening yet.

[222]See a posting on FT Alphaville by Sam Jones *Triple A failure*, 4th July 2008, available at ftalphaville.ft.com/blog/2008/07/04/14317/triple-a-failure.

[223]The statement comes from testimony on the 9th December available at oversight. house.gov/documents/20081209145847.pdf

[224]Bloomberg quotes \$591B in an article of 29th September by Yalman Onaran and Dave Pierson on www.bloomberg.com. The data in the figure following comes from the same source.

[225]The *Shareholder Report on UBS's Write-Downs* (2008), available via www.ubs.com is explicit on this point: 'The UBS funding framework ... facilitated relatively high levels of CDO warehouse activity, as the securities in the warehouse provided positive carry'.

[226]The figures are approximate and the attachment point of the AAA depends on the credit enhancement used as well as the details of the underlying mortgages. Still, the \$300B figure is a reasonable estimate.

[227]Some of the risk was passed on in unfunded form as we discuss in chapter 8, but much of it was *funded* either by banks or in bank-sponsored conduits and SIVs.

[228]Data from Citigroup earnings releases available via www.citigroup.com/citi/fin/ qer.htm. Like most firms, Citi took losses for a variety of reasons during the Crunch: subprime exposure was the most important source, but there were others. See the earnings releases for a detailed breakdown.

[229]The estimate is from FT Alphaville: see ftalphaville.ft.com/blog/2008/08/29/ 15447/merrill-losses-wipe-out-longtime-profits/.

[230]The risk estimate is the average Citigroup 2007 daily 99% as reported in the OCC's *Quarterly Report on Bank Trading and Derivatives Activities* (First Quarter 2008) available at www. occ.treas.gov/ftp/release/2008-74a.pdf. The idea of illustrating the mis-estimation of risk by comparing the average daily VAR with the daily loss comes from the blog econompic data.blogspot.com.

[231]This is a valid concern for two reasons. First, with the almost complete illiquidity of the market, fair values came almost exclusively from models. Different models differed quite significantly in the valuation of the same ABS, so the write-downs for the same position varied from bank to bank, as different banks used different models (with different calibration) for the same security. (The SEC was sufficiently concerned about this that they issued guidance on the valuation of ABS.) Second, some banks held ABS in an accrual rather than a fair value situation, and so it was likely that these banks had not recorded the same level of write-downs as fair value institutions.

# Chapter 7

---

# The Legacy Fails: Fannie, Freddie and the Broker/Dealers in 2008

*On the pedestal, these words appear:*
*My name is Ozymandias, King of Kings,*
*Look on my Works, ye Mighty, and despair!*
*Nothing beside remains. Round the decay*
*Of that colossal Wreck, boundless and bare*
*The lone and level sands stretch far away.*
Percy Shelley

---

## Introduction

When we left the American financial system at the end of chapter 5, we had got as far as the 1970s. The legacy of the New Deal era reforms was still strong:

- Fannie and Freddie were playing an important rôle in the residential mortgage market;

- There was a split between old-style banks on the one hand (regulated by the FED and not active in securities trading) and the broker/dealers on the other (regulated by the SEC, and not active in lending or deposit taking).

What happened next?

## 7.1   The Growth, Distress, and Rescue of Fannie Mae and Freddie Mac

Fannie and Freddie were set up to support residential mortgage lending. After gaining their freedom from Lyndon Johnson, they took up this mission with gusto. They were government sponsored enterprises, which allowed them to issue debt at low credit spreads. They spent the money they raised by buying a lot of mortgages. Then, during the 1980s Fannie and Freddie played a leading part in the invention of MBS. Issuing pass throughs allowed them to acquire more mortgages without worrying about prepayment risk: each GSE could buy mortgages, keep the default risk, but allow the bond markets to fund the mortgages and take the risk of early prepayments. What happened next?

### 7.1.1    Fannie and Freddie Prosper

The business model of private shareholders but government cost of debt is a wonderfully successful one. Fannie Mae's 2001 annual report is entitled *The Dream Decade*, and it certainly was, at least for the two GSEs' shareholders. The two firms' approximate earnings and net assets are shown in figure 7.1[232]. Note that both firms had the risk on more mortgages than are reported under their total assets, since the guaranteed portfolio does not appear on the balance sheet.

| | Fannie Mae | | Freddie Mac | |
|---|---|---|---|---|
| | Net Income | Total Assets | Net Income | Total Assets |
| 1997 | $2.99B | $392B | $1.39B | $195B |
| 1998 | $3.35B | $485B | $1.70B | $321B |
| 1999 | $3.83B | $575B | $2.22B | $386B |
| 2000 | $4.45B | $675B | $2.37B | $459B |
| 2001 | $5.89B | $800B | $3.93B | $617B |

**Figure 7.1**: Fannie and Freddie in the second half of 'The Dream Decade'

These two GSEs display good earnings growth: their assets are rising too. Indeed, assets rose faster than capital. Part of the reason for earnings growth, then, was increasing leverage. On some measures Fannie and Freddie became riskier, to the benefit of their shareholders, but at the expense of increasing risk to the taxpayer. It was perfectly rational for the GSEs to do this: the duty of their boards was to shareholders. What is surprising is that the taxpayers' representatives did not constrain this rising risk.

### 7.1.2    Oversight

Fannie and Freddie are regulated by the Office of Federal Housing Enterprise Oversight or OFHEO. The main restriction on risk taking is provided by that body's capital requirements. Specifically, Fannie and Freddie have to keep enough capital to meet the OFHEO's Risk Based Capital Rules. And during the Boom years, they did meet these requirements: Fannie had 25% more capital than it needed at the end of 2006: Freddie Mac had an even bigger percentage safety margin[233]. However these rules suffer from the same problem as the Ratings Agencies' models: they are based on the idea that there should be enough safety to deal with recently experienced changes in house prices. Since history for the OFHEO starts in 1984, the resulting capital requirements did not prevent Fannie and Freddie from taking more and more risk.

### 7.1.3 Their Own Private Idaho

The growth of subprime posed a threat to the GSEs. Most subprime mortgages were not conforming, so Fannie and Freddie could not immediately buy them. The New York Times reports that Angelo Mozilo, the head of Countrywide Financial, took a robust view of this, telling Fannie Mae that they were 'becoming irrelevant'[234]. The growth of subprime, then, caused considerable tensions within the GSEs. These are revealed in the Waxman Committee hearings of December 2008[235] which illustrate some of the conflicts. The argument within Fannie and Freddie could have gone something like this:

> 'Subprime is not bad lending *per se.*'
>
> 'But mis-priced subprime lending, where the lender is not receiving enough compensation for the risk, is.'
>
> 'We have a duty to support the market. In particular we have a duty to support the low income end of the market. That's subprime.'
>
> 'But we may not be able to properly assess the risks of these loans. Our risk management people are not comfortable with this.'
>
> 'We are losing market share. We have congressionally mandated lending targets to hit, as well. The political consequences of backing away from this market will be severe.'
>
> 'We are so large that we can shape the market. We did it in 1985 – we can do it again. There is no need to capitulate. We should withdraw from the riskiest kinds of lending.'
>
> 'We can't. We need to participate in this market or we won't make enough money.'

Of course this is a fiction. But it does illustrate the dilemma the GSEs faced. They began to increase the proportion of risky loans they purchased, albeit at a slower pace than some urged. For instance Freddie Mac's exposure to loans with a current LTV of over 90% doubled from 2001 to 2007[236]. They increased their purchases of both Alt-A and lower credit quality mortgages too[237], although they were still constrained by the requirement that they could only purchase whole loans if they were conforming.

Fannie and Freddie had another tactic to increase their subprime activities: buying other peoples' MBS. They argued that the purchase of private label MBS was a natural extension of their rôle in buying mortgages. Thus these securities started to appear on both sides of Fannie and Freddie's balance sheet: as liabilities, funding their mortgages; and as investment assets. Thanks to their implicit government guarantee, the Agencies were able to raise funds for these investment purchases rather cheaply. And, best of all, the loans behind these investment MBSs did not need to be conforming.

Fannie and Freddie's portfolios of MBS grew to be astonishingly large: by late 2007 Fannie and Freddie owned $1.4 *trillion* of investments[238]. To put this in context at the time the U.S. national debt was $9 trillion.

### 7.1.4   Fannie and Freddie in the Greenspan Boom

To summarise, Fannie and Freddie had four major rôles prior to the Crunch:

- They bought mortgages from banks;

- They retained the default risk of these loans, and some of the prepayment risk;

- They issued debt, including significant volumes of MBS. These securities passed on the rest of the prepayment risk; and

- Increasingly, they bought private label MBS as an investment.

The strategy of enhancing returns to shareholders by increasing leverage was pursued with vigour. The two GSEs had core capital (as defined by their regulator) of $83.2 billion at the end of 2007. But this supported $5.2 trillion of debt and guarantees, a leverage ratio of 65 to one[239]. Not only was their total risk over half of the U.S. national debt, but the leverage is much higher than a bank would have been permitted. Wachovia, for instance, had a leverage of only 14 to one at the same date[240].

### 7.1.5   Losses Mount

It has been argued that Fannie and Freddie were weakly supervised and strayed from the core mission. Certainly during the Boom years they used their cheap funding to buy MBS backed by pools of mortgages that were rather riskier than their usual investments. By mid 2008, it became clear that the relatively small amounts of capital each GSE had was not enough to cover the losses they were exposed to. Figure 7.2 illustrates the problem[241].

|                      | Fannie Mae | Freddie Mac |
| --- | --- | --- |
| Q1 2007 net income   | $0.8B   | -$0.2B  |
| Q2 2007 net income   | $1.8B   | $0.6B   |
| Q3 2007 net income   | -$1.5B  | -$1.3B  |
| Q4 2007 net income   | -$3.5B  | -$2.6B  |
| Q1 2008 earnings     | -$2.5B  | -$0.4B  |
| Q2 2008 earnings     | -$2.6B  | -$1.0B  |

**Figure 7.2**: Fannie and Freddie as the Credit Crunch bites

This was not a liquidity problem, as suffered by Bear Stearns or Northern Rock. The GSEs could still raise money. This was a *solvency* problem. Investors were concerned that losses were endangering Fannie and Freddie's capital base. Indeed, some were nervous that the situation was worse than was visible in the firms' earnings. After all, private sectors firms with mortgage portfolios rather smaller than that of the GSEs were taking much bigger losses. Fannie and Freddie were not required to mark all of their portfolios of mortgage risk to market, and doubtless their losses in the Crunch would have been higher had they done so. Finally there was a history of accounting issues at both Fannie and Freddie which may have further reduced confidence in their disclosures. In any event, the share prices started to slide. Even more challenging to the two GSEs' business model, their credit spreads started to rise: it became more and more expensive for them to raise new debt. The cost of protecting against their default soared: Fannie's CDS spread, for instance, was teens of basis points during 2006; by late August 2008, it was 350 b.p.s. By this point Fannie's shares had lost over 90% of their value.

### 7.1.6   Conservatorship

At this point, the implicit government guarantee had to be made explicit. House prices were already plummeting, and the GSEs were needed to support the faltering mortgage market. The financial system had racked up hundreds of billions of dollars of losses. Both GSEs had issued trillions of dollars of debt, and this was widely held throughout the financial system. A Fannie or Freddie failure would have posed enormous systemic risk and would almost certainly have caused a global meltdown. Therefore the government had to step in. On 7th September 2008, the Federal Housing Finance Agency placed Fannie Mae and Freddie Mac into *conservatorship*. This is a legal process in which the powers of the GSE's directors were transferred to the FHFA. At the same time, the Treasury recapitalised each GSE. Fannie and Freddie were forced to give the Treasury warrants to buy 79.9% of their common stock in exchange for an agreement that the Treasury would buy sufficient preference shares in each firm to keep them afloat. In effect, then, the U.S. government seized control of both firms and forcibly recapitalised them. Existing shareholders were substantially diluted (although they did not lose everything[242]).

It is interesting to note that once in conservatorship the GSEs took a much more robust attitude to the value of their portfolios. Fannie reported a net loss of $29.4B for the third quarter of 2008: Freddie's loss was $25.3B.

### 7.1.7 Full Circle: The Consequences of Conservatorship and the Future of Fannie and Freddie

The key elements of the conservatorship of Fannie and Freddie, according to the FHFA's director, James Lockhart, are[243]:

- To provide stronger backing to the holders of Fannie- and Freddie-issued debt;

- To permit the GSEs to increase their mortgage exposure, allowing them to continue to support the market. Given that by mid 2008 the risk of over 80% of new mortgages were taken by the GSEs, this move was necessary to avoid a catastrophic decline in the availability of mortgage finance.

- The appointment of new management;

- The suspension of dividends on common stock, conserving earnings; and

- The end of the GSE's extensive political lobbying activities.

Simultaneously the Treasury department provided liquidity to Fannie and Freddie (and the FHLBs) via a new *GSE Credit Facility*. This allowed the GSEs to borrow from the Federal Reserve, pledging MBS as collateral[244].

The conservatorship of Fannie and Freddie was a credit event under standard credit derivatives documentation. Therefore the vast majority of the credit default swaps on the GSEs terminated in September 2008, with the protection buyers having to compensate the protection sellers for the consequences of conservatorship.

Objectively, these consequences were minimal: neither GSE defaulted, and in fact conservatorship *increased* the credit quality of the GSEs' debt. The payments under most CDS were set by an auction process[245]: this allowed protection buyers and protection sellers to agree a price at which the market would clear. Despite a number of Cassandra-like warnings in the days before the auctions, the process passed off smoothly, and the net payments from protection sellers to buyers were indeed small[246].

By the time the subprime problems were evident to all, Fannie and Freddie's market support was needed more than ever. This is why they are being permitted to *increase* their MBS portfolios through 2009, despite being in conservatorship. They are returning to their original rôle of supporting home ownership without concern for their shareholders. But it is unclear whether that artefact of Lyndon Johnson, the shareholder owned but government supported enterprise, will eventually resurface.

### 7.1.8   And finally… The FHLBs

The story of the FHLBs is similar to that of Fannie and Freddie, so we will only touch on it briefly. The Federal Home Loan Bank system was chartered by Congress in 1932 to provide liquidity to thrifts and the housing finance industry[247]. The 12 regional FHLBs lend against a range of collateral including residential and commercial mortgages for periods from overnight to some years. FHLB lending is entirely to financial institutions, specifically to the home loan banks' 8,000 odd members.

The FHLBs have become a more significant source of liquidity to the mortgage market as the Crunch deepened, just like F&F. Advances from the FHLBs grew nearly 60% from the end of 2006 to September 2008[248], and it is highly likely that the value of the collateral backing these advances steadily declined.

The most immediate problem for the FHLBs, though, was their investments. Again echoing Fannie and Freddie, the FHLBs had interpreted their mortgage market support mandate to include buying private label MBS. The reportedly owned $76B face value in September 2008[249], with losses of over $13B. Given that the FHLB's capital was only $57B, it was clear that the FHLB's solvency was seriously challenged. Whether they were well-capitalised at this point is a matter of judgement[250]. Certainly their ability to pay dividends to their members was scant. By early 2009, it seemed that conservatorship, or at least significant draw downs on the Treasury's GSE credit facility[251], were inevitable.

## 7.2    Financial Services Modernisation in the 1990s

The New Deal era left American financial services in several shards. In one were the banks, taking deposits and making loans, overseen by the FED (and the OCC, and the FDIC). In another were the broker/dealers, trading securities, overseen by the SEC. The insurance companies formed another group. By the mid 1990s, this arrangement was looking rather old-fashioned. The buzzword then was *consolidation*. Large European and Japanese financial services conglomerates were already engaged in both banking and capital markets activities. Some of them even included insurance companies within the group. The idea was these 'financial supermarkets' could sell a complete range of retail banking, commercial banking, wealth management and wholesale financial services products, and that this integrated activity would be more profitable than standalone companies.

### 7.2.1    Was Modernisation Necessary?

The failure to update the broad architecture of U.S. financial services since the 1930s had left America ill-positioned for the globalising markets of the 1990s. The rapid growth of the Japanese banks in the 1980s demonstrated that American supremacy in finance was by no means guaranteed: the largest bank in the world was the Bank of Tokyo-Mitsubishi, and European banks such as Credit Suisse and Deutsche Bank were challenging the Americans too. Given that there was no big crisis to encourage prudence, the result was inevitable: *deregulation*.

### 7.2.2    Building an American Supermarket

One decisive moment came in 1998. Salomon Brothers was one of the best regarded broker/dealers in the world. It was active in securities trading and in the rapidly growing derivatives markets of the 1990s. However it was relatively small and – compared with large banks anyway – badly capitalised. That made it a takeover target, and in 1997 the Travelers Group purchased it. Travelers was primarily an insurance company, but it also included a retail brokerage, Smith Barney, so it had every component of a financial services supermarket except banking. This group clearly offered a tempting partner for a pure bank: by combining with Travelers, a commercial bank could immediately acquire all the pieces of a modern, integrated financial services group.

There was only one problem with this idea. It was probably illegal. Glass-Steagall did not permit banks to engage in all these activities, at least in the size necessary to be a global champion[252]. However, the political mood was accommodative towards business. Therefore when Citibank proposed a merger with Travelers in 1998, it was granted a waiver from the provisions of Glass-Steagall. The resulting conglomerate, Citigroup, was the largest financial-services company in the world, just stealing the title from the Bank of Tokyo-Mitsubishi. It had almost $700B of assets, and with a market capitalisation of $135B, it was the world's biggest company[253].

### 7.2.3  Broadening the Brokers

It is worth noting that while American banks were contemplating their moves into broker/dealer territory either by acquiring companies or by building their own capital markets operations, the broker/dealers were also pushing from the other side. There were three main routes here:

- Glass-Steagall did not constrain the foreign activities of the broker/dealers, so they were free to set up subsidiary banks outside the U.S. Typically they picked tax 'efficient' jurisdictions for these banks such as Switzerland or Ireland.

- Glass-Steagall also did not prohibit the broker/dealers from some parts of retail banking such as credit card issuance. Thus in 1997 Morgan Stanley (a leading broker/dealer) merged with Dean Witter (a retail brokerage and credit card issuer).

- Finally, an obscure loophole in U.S. banking regulation allowed the broker/dealers (amongst others) to set up quasi-banks known as *industrial loan corporations*. These ILCs could and did take deposits and make loans[254] but without the constraints of FED supervision.

The trend, then, was for bigger financial services groups with more functions[255].

### 7.2.4  Breaking down Glass-Steagall

Clearly it would be unfair for Citigroup to be allowed to be a special case: the restrictions of Glass-Steagall had to be weakened. Thus in 1999 American financial services legislation was modernised with the passage of the *Gramm-Leach-Bliley Act*. The major provisions of this piece of legislation were[256]:

- To repeal the Glass-Steagall restrictions on banks affiliating with broker/dealers;

- To create a new kind of corporation, the *financial holding company*. These could engage in banking, capital markets activities such as securities underwriting, and insurance;

- To regulate each type of activity *separately*. That is, the State Insurance supervisors would continue to regulate insurance activity, whether carried out as part of a larger group or not; and the SEC would continue to regulate the broker/dealer subsidiaries of banks as well as the broker/dealers themselves; the FED would be an 'umbrella' regulator for financial holding companies.

This last point is key. The Act offered a route for the banks to become supermarkets. But that was not the only route. The broker/dealers were *not* forced to become financial holding companies. Instead they were allowed to continue with their pre-GLB plans as supervised by the SEC.

Thus there were different lead supervisors with different rules and different capital requirements depending on whether the supermarket was a Financial Holding Company or a broker/dealer. Changes in Glass-Steagall were almost certainly inevitable, especially given the international context and the political climate. But the U.S. missed an opportunity to provide a uniform regulatory framework with fewer regulators, fewer types of institutions, and fewer opportunities for regulatory arbitrage.

## 7.3    The End of the Broker/Dealer

Prior to 2004, the SEC regulatory regime for the broker/dealers had long struck some commentators as unfair. In particular, while all large banks had capital requirements for the whole company, broker/dealers were only required to keep capital against the risks in some subsidiaries. There was no consolidated capital requirement of all the market or credit risk in the whole company. This, some banks thought, gave the broker/dealers an unfair advantage. Set against this, though, were the advantages the banks had. They could get a lot of their funding from retail deposits[257], and they had access to the central bank window. By late 2008, the bank model had prevailed, and all five of the large broker/dealers were gone. Two had failed: Bear Stearns and Lehman. Merrill Lynch had been bought; while Morgan Stanley and Goldman Sachs had announced that they were going to convert themselves into banks. To lose one large firm out of five is unfortunately; to lose five definitely smacks of carelessness. Why did things go so badly wrong for the broker/dealers – and, by implication, their regulator the SEC?

### 7.3.1    The Europeans Force Change

The SEC introduced a new capital regime for the largest broker/dealers in August 2004. This allowed the big five to ask for permission to be regulated as *consolidated supervised entities* or CSEs. For the first time, these firms would have minimum capital requirements based on their total risk. The CSE capital regime was long coming: even one of its authors admitted the prior rules were 'considered somewhat rudimentary and antiquated'[258].

The CSE capital requirements were not produced in a vacuum. The European Union adopted the *Financial Conglomerates Directive* in 2004[259]: this provided a strong incentive for the SEC to do something. In particular, the FCD required European supervisors to determine whether a regulated entity 'the parent undertaking of which has its head office outside the Community' was subject to supervision 'equivalent to that' within the EU. Since there was a significant risk that the SEC's capital regime would not be judged equivalent, the SEC had to do something[260].

### 7.3.2    Consolidated Supervised Entities

A broker/dealer became a CSE by applying to the SEC for permission. At the same time, the broker/dealer had to consent to group-wide supervision[261]. The rules for CSEs were based on the capital requirements for banks: they had to be, in order to achieve the necessary equivalence with EU rules. But they were less constraining. The net effect of the CSE regime in practice was to *increase* the leverage of the broker/dealers. Lehman Brothers, for instance, had a leverage ratio of 22 to 1 in 2003: by 2007, it had risen to 30 to 1[262].

It is not surprising that rules developed for the largest banks were not wholly appropriate for the broker/dealers: these firms had a very different balance of risks from an average bank. In particular while many banks are dominated by credit risk – thanks to their loan books – the broker/dealers took significant market risks. Their funding liquidity risk was higher, too, as we saw in the failure of Bear Stearns.

### 7.3.3    The Rise of Lehman Brothers

The history of Lehman Brothers provides an interesting insight into the firms that became CSEs. Lehman was originally set up as a cotton trader in Alabama. The two Lehman brothers starting working together in 1850. When commodities trading centralised in New York, they set up a branch office there. By the end of the nineteenth century, the firm had added equity trading and underwriting to its capabilities.

Lehman survived the Great Depression and grew with the securities markets. By 2000, it was a specialist in bonds and in commercial real estate. It had consolidated its position as a top five broker/dealer, and was well known for a range of capital markets activities. Like a number of other firms, Lehman Brothers enhanced its credit derivatives and securitisation capabilities as the ODM grew during the Boom years. It even bought a subprime mortgage company, BNC Mortgage, to increase the flow of mortgages available for securitisation.

At first it appeared that Lehman's strategy was working. The 2007 Annual Report detailed not just positive earnings, but the highest ever net income for the firm, $4.2B. The firm's return on equity was 21%: this was much higher than most banks could boast. Its net write down for 2007 was small, at $1.9B[263]. Superior risk management appeared to have steered Lehman successfully through the crisis.

Less than a year later, it was bankrupt.

### 7.3.4   Choppy Seas

The Waxman Committee hearing into the causes and effects of the bankruptcy of Lehman Brothers provided some good insights into the firm's strategy in early and mid 2008[264]. Criticising themselves, Lehman's management identified several mistakes. To summarise:

- They originated too many illiquid assets for too long. That is, despite the growing signs of trouble through 2007, the firm continued to structure transactions which created illiquid assets. As the market deteriorated, they could not sell all of these assets.

- The firm did not have enough discipline about capital allocation. In other words, it entered into trades where the potential return was not sufficient to justify the risk being taken.

- It behaved too much like an investor, and not enough like a trader. The slogan for the broker/dealers was traditionally 'we are in the moving business, not the storage business,' meaning that the firm made its money from buying and selling, with little retained risk. The returns available from storage were so tempting during the Boom years that all of the broker/dealers moved away from their traditional model to some degree. But Lehman went too far.

Investors, having noted these issues before the firm itself did, sold the stock. The share price fell through much of 2008. Lehman reported a net loss for the second quarter of the year – $2.8B – and this caused a reappraisal of strategy.

The firm's well-regarded chief financial officer left, and crisis management was stepped up a gear. Some assets were sold, deleveraging Lehman's balance sheet. The firm raised some new capital, and looked into various arrangements to reduce its risk or raise money[265].

## 7.3.5   The Sudden Fall

It was not enough. The key incident came when the news leaked on the 3rd September 2008 that attempts to raise further new capital from Korea Development Bank had failed. Rumours about the Lehman's liquidity position intensified. Those investors who were short – betting that the firm's share price would fall further, or that its credit spread would increase – took the news as confirmation of the correctness of their position[266]. The share price continued to fall, as illustrated in figure 1.8.

As the markets fell, debt investors became more and more concerned about Lehman's viability. The firm found it increasingly difficult to roll its funding. It needed at least $100B a month to finance its assets, and these were almost all sourced from the wholesale market[267]. Thus when debt investors began to decline to refinance the firm, Lehman was in trouble. In an attempt to reassure the markets, Lehman rushed out a third quarter trading statement on 10th September, detailing a $3.9B loss for the quarter. Simultaneously it announced plans to sell more assets including Neuberger Berman, the firm's asset management arm[268]. The market was not convinced: Lehman's credit spread widened further, and it became impossible for the firm to borrow enough money to meet its obligations.

The example of Bear Stearns shows that the end is quick at this point. On the 13th September the FED arranged a meeting to see if Lehman could be saved. Two banks in particular were reported to be interested – Bank of America and Barclays – but neither was willing to rescue Lehman. Lehman Brothers was forced to seek bankruptcy protection. The firm's chapter 11 filing on 15th September 2008 was the largest in American history.

## 7.3.6   The Aftermath

Why did the American authorities let Lehman Brothers go bankrupt? After all, in the earlier and rather similar case of Bear Stearns, the FED had arranged a rescue, albeit a shotgun marriage. One answer is moral hazard. Perhaps the decision was made that for once investors had to bear the consequences of their decisions. And those consequences were certainly acute: the credit derivatives market settlement for Lehman Brothers was 8.625%, implying a loss of more than 95% for debt holders. Equity holders lost everything.

Another possibility is Lehman's balance of business. Bear Stearns was a large player in the MBS market: liquidating its holdings would have further depressed an already illiquid and heavily sold market. Lehman's assets were more diverse. Perhaps the judgement was made that its failure involved less systemic risk, particularly as the market had been prepared for the failure of a large institution for much of 2008.

In any event, the effects of the Lehman bankruptcy were severe. The firm's financial assets are being liquidated to pay creditors, a process which might take ten years or more. Its business was carved up between Barclays (who bought much of Lehman's North American operation, but without the inventory) and Nomura (who bought significant parts of the firms' activities in Europe and the Asia Pacific region). Neuberger Berman was sold to its management, realising far less for creditors than the original estimate of $8B.

The failure of Lehman Brothers was almost as much of a problem for the other broker/dealers as it was for Lehman's creditors. It highlighted the extreme vulnerability of their business model to a change in confidence and hence a wholesale funding run. Bank of America had shown interest in buying Lehman during the FED talks: Merrill Lynch noted this appetite for acquisition, and quickly offered itself for sale. A transaction was quickly arranged with Bank of America paying a significant premium to Merrill's stock price to take control of the firm in a $50B deal. Within two days, two of the four remaining broker/dealers had disappeared.

The spotlight quickly turned onto the other two. Morgan Stanley and Goldman Sachs both applied to the FED for permission to abandon their status as broker/dealers and become banks. By the 22nd September 2008 they had the required permission[269].

## 7.4 Lessons from the Failure of the Broker/dealer Model

The disappearance of a whole class of firm over a short period of time is unprecedented. After Lehman, the broker/dealer model was seen to have failed so badly that even the three surviving large firms decided that their best hope was in banking. Here then we look at what was unique about the broker/dealers and why those characteristics were ultimately fatal.

### 7.4.1  What Distinguished the Broker/Dealers

The broker/dealers were highly profitable in the good times. They had high earnings; their executives were very well paid; and their returns on equity were higher than those of most banks. This made them attractive places to work for risk takers.

Another feature of the broker/dealers was that some of them were relatively immature as public companies. To take two examples, Goldman Sachs was a partnership until 1999, while Lehman Brothers became an independent public company in 1994. Goldman's history is sometimes credited with creating a light-weight yet effective risk management culture: until the IPO, the managers were also largely the owners[270]. But at Lehman concerns were raised by some commentators that the firm's corporate governance was substandard[271]. The problem is that bureaucratic controls and governance make it hard to make money: a fast and adaptive control infrastructure is needed to take advantage of opportunities. When this works well – as seems to be the case at Goldman – then the firm can be highly profitable. But when management convince themselves that they know better than the market, and corporate governance does not provide an effective counterbalance, then the firm can sleep walk over a cliff.

Broker/dealers are inherently riskier than banks because of their funding. A bank can take deposits. Retail deposits provide stable, low cost funding (at least when they are taken in a country with an effective deposit protection regime). The broker/dealers' inability to use deposits to fund their assets was a key vulnerability[272]. It meant that they had to rely on more volatile forms of funding such as repo or commercial paper.

### 7.4.2 Reviewing the CSE Regime

Weak regulation is in some ways worse than no regulation: it can give investors false comfort. The presence of regulators can also be used to justify weaker disclosure. If someone else is watching, do you need to be wary too?

In this context it is worth examining how well the SEC's CSE programme worked. Fortunately we have good information here: at the request of Senator Grassley, the SEC's oversight of the big five broker/dealers was audited after the failure of Bear Stearns. The conclusions of the audit suggest that SEC regulation of the CSEs was far from perfect. In particular the following failings were identified:

- Bear Stearns was in compliance with the CSE program's capital and liquidity requirements. Yet it failed. The same is presumably true of Lehman Brothers. Therefore it seems clear that the requirements were not onerous enough.

- The SEC was aware of numerous red flags prior to Bear Stearns' collapse, including its concentration in MBS and 'numerous' risk management inadequacies. Yet it did not take action. Again, any read over to Lehman at the moment would be pure conjecture, but in the absence of other information, bankruptcy does seem to indicate failings of risk management.

- The SEC's rules prior to the CSE programme limited debt to fifteen times net capital. Once the large five broker/dealers became CSEs, there was no comparable restraint. We have already seen how this allowed Bear Stearns and Lehman Brothers to increase their risk taking.

- The suggestion that the CSE program was a sop to the Europeans rather than an honest attempt to regulate the broker/dealers – perhaps an unworthy suspicion – could be seen to be supported by the fact that the SEC gave permission to four of the five broker/dealers to become CSEs before their inspections were even complete. They then did not monitor the firms' regulatory reporting in a timely manner. This certainly raises concerns about the effectiveness of SEC supervision.

The conclusion is perhaps best left to Senator Grassley. He said that the audit reports[273]

> ... document the failure of regulators at the Securities and Exchange Commission to either make its oversight program work or seek authority from Congress so that it could work. Instead, the SEC Inspector General found that officials responsible for monitoring the safety and soundness of our nation's largest investment banks ignored red flags and risky behaviour. Those institutions are now gone or re-organized and our financial system is in crisis.

The CSE regime was brought to an end on the 26th September 2008. The chairman of the SEC, Christopher Cox, acknowledged even as he closed the programme that it had been 'fundamentally flawed from the beginning', in part because of the voluntary nature of consolidated supervision. Whatever the reasons, this had not been a glorious episode in the history of American financial services regulation[274].

### 7.4.3    An Old Architecture in a New World

The New Deal reforms of the 1930s left America with a regulatory structure which served it well into the 1980s. But by that point, modernisation was needed. Instead of unifying the disparate regulatory frameworks and multiple classes of entity that had been inherited from the pre war era, the U.S. simply loosened regulation.

The result was a system where some types of entities had significant advantages over their competitors. The large banks were allowed to buy securities firms and offer insurance. The broker/dealers avoided consolidated capital requirements for a long time, then were given a capital regime which did little to constrain their leverage. Fannie and Freddie were permitted to use their low cost of funds to the advantage of their shareholders.

The inherent flaws in these business models did not bite until the Credit Crunch arrived. Falling house prices then revealed that taking the risk on a multi-trillion dollar mortgage portfolio in a highly leveraged entity can go awry. Falling confidence revealed that funding multi-hundred billion dollar securities portfolios via repo and CP can be dangerous too.

Fannie and Freddie are now back in public hands. And Glass-Steagall is dead because there are no major firms left on the securities side of the divide. The legacy of the New Deal reforms has proved rather mixed.

## 7.5   Compensating Controls

There are few issues which attract the same intensity of debate as executive compensation. Some think that any attempt to restrain rewards to executives will dilute growth and inhibit innovation. Others see a growing gap between rich and poor as inequitable. Many deplore what they see as payments for failure. This latter view has attracted more and more adherents given the high levels of compensation enjoyed by the managers of the broker/dealers during the Boom years, and the subsequent history of many of the firms that they managed.

A neutral approach to the issue might be to ask what is good for the system. That is, what form of executive compensation in financial services produces the 'best' behaviours (whatever we mean by 'best'). So with that in mind, what are the features of the current system?

### 7.5.1   Incentive Compensation

Suppose you are rational and profit-maximising (in itself a highly questionable assumption[275]). Then if I say that I will pay you 20% of any amount you make over £1,000,000, clearly I am incentivising you to make as much money as possible. You are being paid for performance.

There are a few problems with this idea however. The first is that most individuals do not work alone. They work in a large firm. If nine traders each make £2,000,000 but one loses £20,000,000, then the whole firm loses money. However, according to our incentive compensation rule, we still have to pay out £1,800,000 in bonuses.

Paying bonuses when the whole firm has lost money arises because typically it is not possible to pay a negative bonus. Not paying the traders who made money is unfair to them. But for the firm to pay when it has lost money is unfair to the shareholder. Really it is the trader who lost £20M who should be paying. Of course, for traders who lost many times their total wealth – as must have happened at the firms with big subprime losses – even schemes which clawback previous years' bonuses would not make up the shortfall[276]. Still, negative bonuses have an obvious appeal to the fair-minded.

Another issue arises because many traders do not make much *realised* profit in a year. The majority of their earnings are on a mark to market basis. This means that subsequent moves in the value of their position can eliminate earlier earnings. We are only sure what they have really contributed once the position is closed. And that may take many years, especially if it is derivatives-based or illiquid or both. One solution here may be to pay based on the long term performance of the position: the risk taker would get a stake in the final outcome, but that would not actually be paid in cash until it was known[277].

## 7.5.2   Collective Bargaining

Any good trade unionist knows that the key to collective bargaining is solidarity. If the workers are united, management will have to accede to their demands. Investment banking was one of the last industries where this was the case. There was no union as such, but individuals acted during the Boom years as if there was. The story was always told that high rewards were necessary to retain talented individuals. The idea that the same individuals would gladly work for a tenth of their compensation – and that similarly talented people had indeed worked for much lower rewards only a decade earlier – was unmentionable. Compensation went up and up, not just absolutely but also as a percentage of the average wage[278].

Compensation committees were supposed to protect shareholders' interests. In fact these bodies were often composed of individuals with a stake in the outcome, either directly or indirectly. It was rare indeed for a recommendation to be made of less than the industry average. If you keep on paying people more than the average, the average goes up.

In practice shareholders have little say in executive compensation. One has to wonder if executives have arranged it that way for their own benefit.

There is no legal requirement for shareholders to even have a vote on compensation in the U.S., for instance[279]; while in the UK, voting against compensation recommendations seems to be a nuclear option which shareholders only exercise if they are extremely displeased with the Board. All of this meant that in practice there was little restraint on executive compensation: when times were good, investment bankers were paid very well indeed.

### 7.5.3   Incentives

The high levels of incentive compensation in financial institutions were a real cost to shareholders. On some measures the fraction of company earnings paid to senior executives doubled between 1993 and 2003[280]. But did it work? Do higher incentives produce better performance?

Larger incentives certainly encourage higher earnings to be reported. One issue is that these earnings may not represent true realised gains for the company, either because mark to market values can fall or because the earnings were largely generated through the use of techniques which produce flattering income statements for a short period (like aggressive accounting[281]).

Another problem is that comparatively high earnings in financial services encourage the avaricious to become bankers. Is it really in society's interest that most of the smart, greedy people work in a single industry[282]?

If we accept the idea that individuals are rational and profit-maximising then compensation policies are the key tool in encouraging the behaviours we want. On this basis the compensation strategies of the Boom years clearly have not worked since we did not get behaviours that were in the interests of shareholders or society. Instead individuals have too often been paid to fail[283]. Vast amounts were paid out in 2005, 2006 and 2007 for actions which ultimately destroyed rather than enhanced shareholder value.

### 7.5.4   Compensation in an Era of State Capitalism

Senior executives at the largest American financial services companies took home total compensation of tens of millions of dollars each in 2007. Fifty million dollars was not an impossible amount for one man to earn in a year[284]. While the firms concerned were entirely private, one could argue that that is a matter for shareholders alone. But now they are not: most of these firms have received multi-billion dollar injections of taxpayers' money. What should the taxpayer do?

It seems that there are three important questions:

- What manner of compensation scheme encourages the behaviours that the owners of a company want to promote and discourages those that they do not favour?

- Specifically, given that the state in many cases owns a substantial stake in the banks, what compensation scheme for bank employees is to the benefit of society?

- How can the overall level of compensation be constrained given the powerful incentives for it to grow and become less performance-related[285]?

This is a difficult problem. But perhaps framing it in these terms may make it easier to address. And it should be addressed quickly. There is a once-in-a-generation opportunity to change the way financial services workers are paid, not least because in 2009 it is likely that there will not be much money made in financial services: principles are often easier to agree when there is not too much immediately at stake. Delaying compensation payments so that shareholders are really sure that the money has been made before any bonuses are paid is clearly part of the answer: disclosure of who gets what[286] and reducing overall levels of compensation may have a rôle in the solution too. But whatever firms and their owners try to do, it won't be easy.

---

# Notes

[232] Annual reports and various other data for Fannie Mae are available via www.fanniemae.com: see also www.freddiemac.com. The net income figures are the reported 'Net Income available for common stockholders' from the firms' annual reports and earnings releases. Care is needed however as both companies identified what the OFHEO calls 'significant errors in accounting policies' and restated their earnings. Moreover both companies report both GAAP compliant and non-GAAP earnings, so the real situation can be difficult to perceive through the various forms of accounting.

[233] See the OFHEO Annual Report to Congress 2007, available at www.ofheo.gov/media/annualreports/OFHEOReporttoCongress07.pdf. The risk based capital rules are available at www.ofheo.gov/Media/Archive/docs/regs/rbcfinalamend21303.pdf.

[234] See Charles Duhigg's article *Pressured to Take More Risk, Fannie Reached Tipping Point*, 4th October 2008, available at www.nytimes.com/2008/10/05/business/05fannie.html.

[235] Documentation of the hearings can be found at oversight.house.gov/story.asp?ID=2252. The Fannie Mae presentation *Single Family Business Guaranty Business: Facing Strategy Crossroads* is particularly insightful.

[236]Table 5 of the 2001 annual report gives Freddie's exposure to loans with an estimated current LTV above 90% as 5% of the total portfolio: by 2007's annual report, the comparable figure from table 46 is 10%.

[237]See in particular the testimony of Edward Pinto on this point, which suggests that the data in the previous note underestimates the growth of low quality loans in the GSEs' portfolios.

[238]See *Understanding the Dilemma of Fannie and Freddie*, The Republican Caucus Committee on the Budget July 2008 available at `www.house.gov/budget_republicans/press/2007/pr20080717banks.pdf`.

[239]The figures here are from an article in the Economist, *End of illusions*, 17th July 2008, available at `www.economist.com/finance/displaystory.cfm?story_id=11751139`.

[240]This is based on Wachovia's Tier 1 ratio of 7.35% as reported in the 2007 annual report available via `www.wachovia.com`. To be fair, the two ratios are not directly comparable, due to the differences between Bank and GSE regulation. Nevertheless these differences are not so considerable that one cannot reasonably conclude that Fannie and Freddie were considerably riskier than they would have been were they to have been regulated in the same framework as U.S. banks.

[241]The data is from Fannie Mae and Freddie Mac investor relations publications, and the OFHEO. Sometimes various sources do not appear to be wholly consistent, perhaps reflecting later accounting restatements.

[242]There is an argument that moral hazard is increased by leaving anything for ordinary shareholders. Still, shareholders suffered losses well in excess of 95% from the highs. Moreover, by keeping the government's share below 80%, it was possible to argue that the U.S. government debts should not include that of Fannie and Freddie. Given that the two GSEs owed over five trillion dollars, this accounting legerdemain was important.

[243]See *Statement of The Honorable James B. Lockhart III, Before the House Committee on Financial Services on the Conservatorship of Fannie Mae and Freddie Mac*, 25th September, 2008 available via `www.ofheo.gov`.

[244]See the U.S. Treasury Department's *Fact Sheet: GSE Credit Facility* (2008) available at `www.treas.gov/press/releases/reports/gsecf_factsheet_090708.pdf`.

[245]See ISDA's *Plain English Summary of the Auction Methodology in the 2008 Fannie Mae and Freddie Mac CDS Protocol*, available at `www.isda.org/2008fmfmcdsprot/docs/Fannie-Freddie-Plain-English-Summary.pdf`.

[246]An interesting small anomaly in the process was that the settlement amount on each GSEs' subordinated debt was *lower* than that on its senior debt. In a real bankruptcy this cannot happen as the senior is in front of the sub debt in the order of payment. In an auction to determine where the CDS market clears, however, it can happen as different parties are bidding on the different priorities of debt and it is hard to arbitrage between the two settlement prices. This is especially so since the most speculative activity before conservatorship had focussed on the GSEs' sub debt, and so there was a significant market in these default swaps, with the senior being less well-traded in the months before conservatorship.

[247]For more on the FHLBs, see Mark Flannery and W. Scott Frame's The Federal Home Loan Bank System: The "Other" Housing GSE, Federal Reserve Bank of Atlanta (2006) available via `www.frbatlanta.org/filelegacydocs/erq306_frame.pdf`.

[248]The data comes from the FHL Banks Office of Finance: advances increased from $640B at the end of 2006 to $1.011T at the end of the third quarter of 2008. See `www.fhlb-of.com/specialinterest/financialframe2.html` for more details.

[249] See Jody Shenn's Bloomberg story of 9th January 2008, *FHLBs May Fall Below Capital Minimums, Moody's Says* available at `www.bloomberg.com/apps/news?pid=20601087&sid=aeB5GL6uSr3A`. The FHLB Office of Finance data previously referred to is also useful here.

[250] The judgement turns on the decision as to whether the losses are 'other than temporary impairments'.

[251] It is notable that Treasury secretary Paulson's statement at the time of the Fannie and Freddie bailout refers to the GSEs in general, rather than just the two entities that were put into conservatorship. The Treasury lending facility established then is available for Fannie, Freddie, *and the FHLBs*. See *statement by Secretary Henry M. Paulson, Jr. on Treasury and Federal Housing Finance Agency Action to Protect Financial Markets and Taxpayers* (7th September 2008) available via `www.ustreas.gov/press/releases/hp1129.htm`.

[252] There had been various cracks in Glass-Steagall prior to 1998. For instance starting in 1987 the FED had permitted banks to set up securities dealing subsidiaries providing they were not too big: see Federal Reserve Bank of San Francisco Economic Letter *The Gramm-Leach-Bliley Act and Financial Integration*, 31st March 2000, available at `www.frbsf.org/econrsrch/wklyltr/2000/el2000-10.html`.

[253] See the International Herald Tribune's article *Citicorp and Travelers Plan to Merge in Record $70 Billion Deal : A New No. 1*, 7th April 1998, available at `www.iht.com/articles/1998/04/07/citi.t.php`.

[254] For a further discussion on ILCs, see Donald Kohn's testimony before the U.S. House of Representatives Committee on Financial Services, *Industrial loan companies*, 25th April 2007 available at `www.federalreserve.gov/newsevents/Testimony/Kohn20 070425a.htm`.

[255] These groups were not just too big to fail, increasing systemic risk. They were also, as Terry Jorde of Independent Community Bankers of America has pointed out, too big to punish. (See his testimony *Examination of the Gramm-Leach-Bliley Act Five Years After Its Passage* before the Senate Committee on Banking, Housing, and Urban Affairs, 13th July 2004 available at `www.icba.org/advocacy/testimonydetail.cfm?ItemNumber=524&sn.ItemNumber=1699`.) Citing several examples of fines and other regulatory interventions, Jorde says, 'While some of the fines and settlements may seem large to the average person, they are tiny when applied to these companies.'

[256] See *Financial Services Modernization Act Summary of Provisions* from the U.S. Senate available at `banking.senate.gov/conf/grmleach.htm`.

[257] The broker/dealers managed to do this too to some extent via off-shore banks or their ILCs, but it was never as important a funding source for them as it was for the banks.

[258] The quote is from the SEC Historical Society Interview with Annette Nazareth, 4th November 2005, transcript available at `www.sechistorical.org/collection/oralHistories/interviews/nazareth/nazareth110405Transcript.pdf`.

[259] See *Directive 2002/87/EC of the European Parliament and of the Council* available at `eurlex.europa.eu/pri/en/oj/dat/2003/l_035/l_03520030211en00010027.pdf`.

[260] The consequences for the European subsidiaries of the broker/dealers had the SEC's supervision not been judged equivalent might have been severe: as a lobbying document at the time puts it, the alternatives were to 'submit to a U.S. regulator deemed equivalent' (i.e. for the broker/dealers to abandon the SEC as their regulator), to 'submit to EU consolidated supervision' or to 'create [and capitalise] an EU holding company for European activities so as to build firewalls between the EU firms and non-EU affiliates.' This, the document acknowledged, would 'drastically alter' how the broker/dealers conducted their business. See the Se-

curities Industry and Financial Markets Association briefing note on the Financial Conglomerates Directive available at `archives2.sifma.org/international/html/financial_conglomerates.html`.

[261] The consent being necessary because the SEC had no legal power to compel such regulation. For further details of the CSE regime, see the SEC's *Alternative Net Capital Requirements for Broker-Dealers That Are Part of Consolidated Supervised Entities* available via `www.sec.gov/rules/final/34-49830.htm`.

[262] In its 10-Q filing with the SEC for 30th November 2007, Lehman reported $668B of liabilities versus $22.5B of stockholder's equity: the comparable figures for 2003 were $297B and $13.2B. The trends are similar for the other four firms: see Appendix IX of the Office of the Inspector General's report *SEC's Oversight of Bear Stearns and Related Entities: The Consolidated Supervised Entity Program*, available at `finance.senate.gov/press/Gpress/2008/prg092608h.pdf`.

[263] The figures come from the 2007 Lehman Brothers Annual Report. See in particular page 47 for the write-downs for the year ending 30th November 2007.

[264] The documents for these hearings and full transcripts are available via `oversight.house.gov/story.asp?ID=2208`.

[265] One was the creation of a 'spinco' to take tens of billions of dollars of real estate assets off the firm's balance sheet. For more details, see *Lehman may shift $32B of mortgage assets to 'bad bank'*, Bloomberg news story, 4th September 2008.

[266] A prominent example of an investor who had been short Lehman for a while, and who was not afraid to say why, was the hedge fund manager David Einhorn. At the end of 2007 Einhorn discussed Lehman's weaknesses at the 3rd Annual New York Value Investing Congress. In particular he cast doubt on the firm's transparency regarding its write-downs, suggesting that it may have been following a similar strategy to its behaviour in the 1998 LTCM crisis when it had exposure but did not aggressively mark down its book. That strategy worked then, but Einhorn suggested the Crunch was different, not least because Lehman had more significant exposure to structured finance, with securitisation accounting for 15% of its income. Assets originating in this business were stuck on Lehman's balance sheet, Einhorn suggested. Finally he cast doubt on gains arising from Lehman's widening credit spread causing a mark-down of liabilities under FAS 159. See Marcelo Lima's article on Einhorn's presentation available at `blog.valueinvestingcongress.com/2007/12/16/david-einhorn-at-the-3rd-annual-new-york-value-investing-congress-reported-by-marcelo-lima/`.

[267] The figure comes from *Diamond and Kashyap on the Recent Financial Upheavals*, New York Times December 14th 2008.

[268] See the Financial Times, *Lehman Brothers loses $3.9bn in third quarter*, 10th September 2008, available at `us.ft.com/ftgateway/superpage.ft?news_id=fto091020080901009557`.

[269] Technically, they became bank holding companies. See Federal Reserve Press Release 22nd September 2008 available at `www.federalreserve.gov/newsevents/press/orders/20080922a.htm`.

[270] For the case in favour of the partnership culture at Goldman Sachs see Charles Ellis' *The Partnership: The Making of Goldman Sachs* (2008).

[271] See the testimony of Nell Minow at the Waxman Committee hearings, 6th October 2008.

[272] Some of the broker/dealers did take deposits via bank or ILC subsidiaries, but typically regulatory restrictions on the flow of funds outside the deposit taker meant that it was difficult to use these funds to support the rest of the broker/dealer's business.

[273] Senator Grassley's comments in full together with the audit reports are available on the web page *Senator Grassley comments on the report of Inspector General regarding investment bank oversight* at `grassley.senate.gov/news/Article.cfm?customel_dataPageID_1502=17404`.

[274] Cox's press release *The End of Consolidated Supervised Entities Program* is available at `www.sec.gov/news/press/2008/2008-230.htm`. For a further discussion of the CSE regime, see John Coffee's *Analyzing the Credit Crisis: Was the SEC Missing in Action?*, New York Law Journal, 5th December 2008 available at `www.law.com/jsp/ihc/PubArticleIHC.jsp?id=1202426495544`.

[275] For a survey of research here, see Livio Stracca's *Behavioural finance and aggregate market behaviour* (2002). Burton Malkiel's *The Efficient Market Hypothesis and Its Critics* (Journal of Economic Perspectives Volume 17, Number 1, 2003) is also highly relevant here.

[276] UBS has recently introduced such a scheme whereby latter losses can decrease the size of earlier bonuses. It is known as a malus: see `www.ubs.com/1/e/media_overview/media_global/releases.html?newsId=158103` for more details.

[277] Credit Suisse adopted a scheme with elements of this idea in 2008, paying traders based on the long term performance of the firm's portfolio of (mostly illiquid) assets. See `blogs.wsj.com/deals/2008/12/18/some-credit-suisse-bankers-livid-over-new-bonus-plan` for more details.

[278] Research by the House Committee on Financial Services found that the average large-company CEO received roughly 140 times the pay of an average worker; in 2003, the ratio was up to 500 to 1. See `frwebgate.access.gpo.gov/cgi-bin/getdoc.cgi?dbname=110_house_hearings&docid=f:35402.pdf`.

[279] This may change if HR 1257 passes into law.

[280] See Lucian Bebchuk and Yaniv Grinstein's *The Growth in Executive Pay* (2005) available at `papers.ssrn.com/sol3/papers.cfm?abstract_id=648682`. If we take financial services compensation rather than the broader economy as studied by Bebchuk and Grinstein, pay grew even faster.

[281] A little derivatives thinking may be insightful here. Incentive compensation is similar to a call option on the P/L. One way to increase the value of a call is to increase the volatility of the underlying. Therefore by paying bonuses, we encourage individuals to *increase* earnings volatility. The situation for managers is even worse. The manager has a call on the basket of P/Ls of the people who work for him or her. One way to increase the value of a basket call option is to increase the correlation between the components of the basket. Therefore bonuses for managers amount to an incentive *not to diversify*. Finally notice that if traders are paid based on their return for unit risk measure, then they are incentivised to take risk that is not captured in the risk measure. Typically this means that they are encouraged to take tail risk, especially as the premium for writing far out of the money options (or default swaps) usually goes straight to their P/L. This in turn makes firms' earnings distributions less normal, and hence standard risk measures less useful.

[282] Thomas Philippon and Ariell Reshef, in *Skill Biased Financial Development: Education, Wages and Occupations in the U.S. Financial Sector* (2007) found that: 'Overall, the share of talented individuals hired by the financial sector has increased very significantly over the past

three decades'. They also 'document an increase in the wages paid to workers in the Financial sector': high salaries are not just for the CEOs, they are also awarded to many less well-
promoted individuals. See also their paper *Wages and Human Capital in the U.S. Financial
Industry: 1909-2006* available at `pages.stern.nyu.edu/~tphilipp/papers/pr_rev15.`
`pdf`.

[283] One might think in particular here of CEOs that presided over the failure of their firms
yet personally made hundreds of millions. It was in this context that Dick Fuld of Lehman
Brothers won the Financial Times' 'Overpaid CEO Award' in 2008.

[284] Lloyd Blankfein, president and chief executive of Goldman Sachs, took home over $53M
in total compensation in 2007 according to his firm's filings.

[285] As Bebchuk and Fried put it, 'managers wield substantial influence over their own pay
arrangements, and they have an interest in reducing the saliency of the amount of their pay and
the extent to which that pay is de-coupled from managers' performance.' See their *Executive
Compensation as an Agency Problem*, CEPR Discussion Paper No. 3961 (July 2003) available at
`www.law.harvard.edu/faculty/bebchuk/pdfs/2003.Bebchuk-Fried.Executive.`
`Compensation.pdf`.

[286] In *Who Killed Katie Couric? And Other Tales from the World of Executive Compensation
Reform* (Fordham Law Review 2008) Kenneth Rosen discusses some of the issues around disclosure. For my taste he seems too pessimistic though. Shareholders have an obvious interest
in who gets money from their company, and requiring disclosure of the identify and total
compensation of any employee who gets, say, at least 50% of what the CEO is paid would seem
reasonable.

# Chapter 8

---

# Structured Finance

> *Ah, to build, to build!*
> *That is the noblest art of all the arts.*
> Henry Longfellow

---

## Introduction

There are few areas of the markets that have attracted the same opprobrium as structured finance. Commonplace, readily understood transactions – mortgages – caused the Crunch. The very ubiquity of the mortgage meant that it was not a good scapegoat. Instead, some commentators turned to something less well known and so easier to demonise: structured finance.

So what is this apparently blame-worthy business? The term *structured finance* refers broadly to any transaction which transfers credit risk in a form other than a simple corporate bond or loan. There are three major classes of structured finance transaction:

1. Credit Derivatives;

2. ABS; and

3. Off balance sheet financing.

In the following sections we look at each of these classes. *Credit derivatives* allow the risk of default of a debtor to be passed on without the debt being sold. They are a key tool in moving risk around the financial system, so we look at how they work, and compare them with another form of risk transfer, insurance. Failure to understand the controls required around and the risks of trading derivatives caused severe difficulties to one of the world's largest insurers, AIG. We look at AIG's Crunch and how the Federal Reserve came to lend the firm over a hundred billion dollars.

*Off balance sheet financing* refers to ways of raising money to buy assets and getting the benefit from them without these means appearing in your accounts. These techniques were extensively used during the Greenspan Boom to enhance the earnings of financial institutions. We look at what firms did, how that went wrong, and how structured finance made some firms more vulnerable to the Crunch.

## 8.1 Credit Derivatives

A credit derivative is somewhat similar to a guarantee. It involves two parties. The *protection seller* receives a fee from the *protection buyer*. In exchange for this, the protection seller undertakes to compensate the protection buyer if there is a *credit event* on a *reference credit* or credits during some fixed period.

To make things simple, let's suppose that the reference credit is (a bond issued by) Ford Motor Company, and the credit event is bankruptcy. Then in a one year *credit default swap*, the protection buyer would pay a premium, and in exchange they would be compensated if Ford were to file for bankruptcy during a specified one year period.

The usual credit events in a credit derivative include bankruptcy and failure to pay (so that even if a company is not formally bankrupt under the law of its home country, but it is not paying coupons on its debt, then the credit derivative is nonetheless triggered)[287].

Many credit derivatives are *physically settled*: that is, if there is a credit event, the protection buyer can deliver the reference credit to the protection seller and receive par[288].

### 8.1.1   *Specifying a Credit Default Swap*

When a credit derivative is traded, the two parties agree:

- How long the protection is for;
- What the reference credit is;
- What the credit events are;
- What the notional is, that what the face value of obligations of the reference credit that are protected;
- How the credit derivative is settled if a credit event happens.

### 8.1.2   *Separating Default Risk From Funding*

Another way of thinking about a credit default swap is to suppose that I own a bond and that I buy a CDS on the same notional which references that bond. Then if the bond defaults, I can deliver it to the protection seller and receive the face value of the bond: while if it does not default, then the bond issuer will give me the face amount when it is due. Provided the protection seller does their job, then, a credit event does not cause a big loss[289]. Therefore the fee on the CDS – known as the *credit default swap spread* – should be close to the credit spread of the bond, in that both of them represent compensation for the risk of default. However, as we saw in section 3.4.3, credit spreads represent more than the compensation for default alone, so the CDS spread will not be precisely equal to the credit spread[290].

Note that one can trade CDS without having bonds, either buying protection in the hope of a windfall gain on default, or selling protection in the hope of receiving premium income without having to pay out.

The position whereby a bond is held and CDS protection on that bond is purchased is known as a *negative basis trade*. Here funding risk – buying a bond and finding the cash to pay for it – has been separated from taking credit event risk on the bond, as illustrated in figure 8.1. This ability to separate risk taking from funding is one of the key features of the derivatives markets. It allows those parties whose cost of funds is high to take risk without having to pay funding. The instrument is instead funded by a low cost of funds firm such as a large bank which rents its balance sheet out. The *unfunded* risk taker will be willing to pay such fees provided there are sufficient advantages to this strategy.

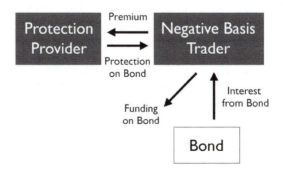

**Figure 8.1**: Negative Basis Trading

### 8.1.3 The Pros and Cons of Exchanges

Credit derivatives are almost always negotiated privately between the protection seller and the protection buyer. No stock exchange intervenes. Thus the market is said to be *over the counter* or OTC. There is considerable price transparency in much of the CDS market: various participants advertise the premiums at which they might be willing to buy or sell CDS protection on a wide range of reference credits. However the market does not have the compulsory trade reporting which is a feature of exchange traded securities.

Notice that merely moving CDS trading onto an exchange does not guarantee liquidity: there are plenty of illiquid securities traded on exchange, and the current price at which one of these securities might be tradable is not magically discoverable. However at least where there is liquidity, the exchange price is likely to be accurate: for CDS, an advertised CDS spread may or may not represent a real trading opportunity.

One risk that is typically larger for derivatives contracts than for securities concerns how long each counterparty is exposed to the performance of the other. When I agree to buy a security, I pay the money and receive the security soon thereafter. A typical settlement period for a security is three days, so if I agree to purchase a Ford bond today, I will actually own it three days later. Therefore I am only exposed to the performance of the seller for a short period of time. It is possible that they can take my money then default before sending me my bond, but it is unlikely. Furthermore, many exchanges remove this risk entirely by providing a *central counterparty*: rather than me facing the seller directly, we both face a central body, and if either of us fail, this body steps in and performs in our place. Thus many securities markets have rather little *counterparty risk*.

This principle is also used in exchange traded derivatives: here the central counterparty guarantees the performance of market participants during the entire life of the transaction. The risk to the central counterparty is reduced by the use of *margin*: when I first trade a derivative, I have to pledge an amount of cash intended to cover the risk that the instrument might turn against me. If my derivative positions fall in value, I have to pledge more cash to the central counterparty to cover the difference, so that they always have a margin of safety; similarly if my positions increase in value, the central counterparty gives me some of my cash back.

Central counterparties are usually high credit quality thanks both to access to significant capital and to the fact that they mitigate risk via margin arrangements. If a large market participant does fail, then they may suffer some losses, but the margin they keep should reduce this to an insignificant sum.

Many OTC markets do not enjoy a central counterparty. Thus for instance if I had purchased one year CDS protection on Ford from Lehman Brothers in January 2008, that protection would have lost effectiveness in

September 2008 with Lehman's bankruptcy as Lehman could no longer perform at that point[291]. Counterparty risk is therefore a major issue in OTC markets.

### 8.1.4  Counterparty Risk

The credit derivatives market is huge: one estimate is that there are over fifty trillion dollars of credit derivatives outstanding[292]. Nevertheless, counterparty risk does not represent as large a threat to financial stability as it may first appear for several reasons:

- Many OTC derivatives counterparties agree to post margin, so the risk of counterparty non-performance is limited to the difference between the value of the exposure and the value of the margin. (Margin can sometimes be pledged in the form of bonds rather than cash. In this instance, confusingly, it is sometimes known as collateral.)

- Counterparty risk is highest by far if there is both a credit event on the reference entity and non-performance of the protection seller in a short period of time. If the former happens, a claim is swiftly made on the protection seller, so the period of maximum risk is short.

- Most dealers hedge their exposure to CDS, so the fifty trillion figure represents a gross rather than a net amount. When a default happens, many market participants will receive roughly equal amounts on protection they have bought to that they pay out on protection sold. The gross amount of money changing hands may be large, but the net profit or loss for most market participants is small.

Perhaps for these reasons the credit derivatives market has worked smoothly during the Crunch. There have been failures of large credit derivatives dealers, including Bear Stearns and Lehman Brothers. Frequently-referenced credits like Fannie Mae have failed too. But the Armageddon created by credit derivatives which had been predicted by some has, thus far at least, not been visited upon us.

### 8.1.5  The Beneficial Effects of Credit Derivatives

Credit derivatives allow firms to hedge. Consider a bank dealing with a large corporate customer. The paradox of banking is that the better the bank knows the client and the better their relationship, the more business they will be able to do; but this business generates risk, and the bank accumulates a more and more concentrated exposure. The ability to buy protection in the CDS market allows banks to hedge this exposure, continuing to serve the client but passing their risk on to third parties. Risk is spread throughout the financial system, reducing the impact of a loss on a single party.

The third parties benefit too. Just as ABS allow them to obtain exposure to risk that they could not easily originate, so CDS also permit them to take risk in a precisely tailored way on a reference credit which may not be available in the bond market. Thus CDS help both to distribute risk around the financial system – lowering costs – and to give better visibility of the price of risk than would be available were all of it to be retained by the originator[293].

### 8.1.6   Risks Created By Credit Derivatives

Credit derivatives do not just pass risk on: they also create it. Consider a bank which has made a loan: it has the risk of non-performance of the loan. Then it buys credit derivatives protection. The protection seller now has that loan non-performance risk. But the protection buyer has newly-created counterparty risk on the performance of the protection seller.

Suppose we set up a company just to write credit derivatives, a *credit derivatives product company* or CDPC. In risk terms this firm would be rather like a bank: it would take credit risk, and support that risk with equity capital. There would be nothing to fund[294], so deposits are not needed: this is pure unfunded risk taking.

At first sight, the only thing which determines our success seems to be whether the CDS premiums received are greater than the losses incurred due to credit events.

But there is another risk: this comes from margin. If the CDS spreads of the companies we have written protection on increase, the CDS we have written become more valuable to the holders. They therefore demand more margin from us. This creates a need for cash *regardless of whether the fundamental credit worthiness of the companies has changed*. If our CDPC cannot meet those margin calls because it cannot raise the cash, it defaults.

Defaults of this kind are known as *credit support defaults*. (The term comes from the fact that the document which sets out the requirements for posting margin is the *credit support annex* to the agreement between the parties[295].)

A writer of CDS protection, then, runs three main risks:

1.    The risk of loss caused by claims on the CDS after credit events;

2.  The risk of changes in value of the contract without a credit event; and

3.  Liquidity risk on any margin requirements generated by changes in value[296].

### 8.1.7  Portfolio Credit Default Swaps

Credit default swaps transfer risk on a single bond. It is also possible to transfer the risk of many obligators at once. A default swap can be written on a collection of obligations: if a credit event happens on any of them, there is a payout. This is known as a *portfolio credit default swap*[297].

There are various forms of these portfolio credit derivatives. The most important are ones that provide protection on all the assets of a CDO, or on a tranche of CDO risk. Thus for instance there might be 100 corporate bonds in a CDO, each with a notional of $10M. The *junior portfolio default swap* on this CDO might pay out on the first $40M of losses but no more. Similarly, a mezzanine swap on the next $50M; and a senior swap on the final $910M. These three swaps together protect the whole $1B of the CDO.

The ability to write portfolio CDS allows us to structure CDOs in a much cleaner fashion. In an ordinary CDO, we cannot hope to find 100 corporate bonds of precisely five years maturity. Moreover most bonds are typically fixed rate, so there is interest rate risk if the issued tranches are floating rate[298]. Finally the bonds will pay at different times, giving us the operational burden of gathering up lots of different cashflows and investing them until we have to pay a coupon on the issued tranches.

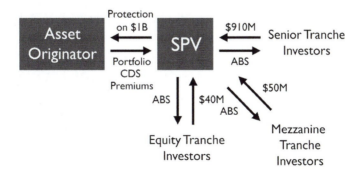

**Figure 8.2**: A synthetic CDO

A synthetic CDO, as in figure 8.2, resolves these issues[299]. Here a SPV takes risk by writing a portfolio default swap rather than buying assets. The default swap can be exactly the desired maturity; premium on the default swap can be paid a day or two before the coupons have to be paid on the tranches, so there are no cash management issues[300]; and there is no interest rate risk.

Another advantage with the synthetic structure is that some loans cannot be transferred without the permission of the borrower. Banks are reluctant to tell their best customers that they do not want to take risk on them. But with credit derivatives, a bank can perform a synthetic securitisation and move risk entirely privately and without reputational damage.

### 8.1.8    The Delta Hedging in Structured Finance

The delta hedging of equity, FX, and interest rate derivatives has been standard for many years. As we discussed in section 6.3.1, delta hedging is often a reasonably effective process for eliminating risk[301].

For some structured finance credit instruments, however, the assumption that delta hedging eliminates risk is more questionable. Consider the AAA-rated tranche of a securitisation of MBS passthroughs. One could either use a model to determine the delta of the tranche with respect to the underlying pass throughs or empirically measure how the spread of the tranche varies with the spreads of the underlying assets. On either basis we can define a *credit delta* of the tranche with respect to each underlying and hedge the tranche with a position in the underlying pass throughs. However if the pass throughs are subject to large jumps in their spread without trading being possible, then the resulting 'hedged' position can in fact jump in value. A risk model based on small movements in spread[302] will show it as low risk. But larger moves in credit spreads can generate significant losses. (In the world of figure 6.7, the kind of jump we are talking about would correspond with a stock price move from $9 up to $12 or down to $6.)

We have to be careful, then. Bought CDS protection might well offer a good hedge against a fixed bond or loan position. But we cannot safely make the assumption that the credit market will always be liquid, or that credit spreads will always move by small amounts.

## 8.2   ABS in Structured Finance

We have met the basic ideas of securitisation, tranching, and the issuance of ABS earlier. Structured finance developed using techniques gleaned from the early GSE pass through and private label MBS deals. Securitisation with tranching became widespread. Then, in the 2000s, deals became more complex. The range of collateral that could be securitised increased; the forms of credit enhancement multiplied and became more sophisticated; and tranching structures became ever more intricate.

This growth was created by investor demand. You cannot sell a complicated highly structured security if no one wants to buy it. Investors were faced with collapsing yields on conventional instruments: they wanted to buy securities that offered a higher return. One form of ABS, in particular, was created to fill the demand.

### 8.2.1 CDOs of ABS

An ordinary CDO uses corporate bonds as collateral to issue tranches. A CDO of ABS, in contrast, uses ABS as collateral to issue tranches. Thus we have a number of first level securitisations which produce ABS, and these ABS in turn go into a second level securitisation as collateral, producing tranches backed by tranches backed by assets[303].

Three features stand out in this process. First, CDOs of ABS can give enough of a gloss that individually unattractive assets can be sold as a diversified package. The more complicated the securitisation is, the more buyers have to rely on ratings or modelling supplied by the structuring bank, and the less due diligence they can do on the underlying assets. Thus subprime mortgages were originally prettified by securitisation and credit enhancement, and if the resulting ABS could not be sold, they were put into a CDO[304] along with a leavening of other types of ABS, in the hope that the diversified package would be saleable.

There was only the pretence of diversification in this process however: 2005 and 2006 vintage CDOs often had sixty percent or more non-prime collateral[305].

Second, a commoditisation takes place. Investors are encouraged to treat MBS of a given rating as uniform objects. CDOs of ABS are sold based on the rating of the collateral, with scant details of the underlying mortgages being available. The investor might be told an average FICO and LTV, but little else. This climate of 'look at the coupon, look at the rating, then buy it' discouraged any vestigial due diligence from structured finance risk takers.

Third, investors may get false comfort from owning senior tranches. In many cases the junior tranche of CDOs of ABS was owned by the deal manager, and their fees were often larger than the value of the tranche[306]. They did not buy the tranche as disinterested investors. The mezz was often bought by other CDOs: again, this was not bought by a third party who had done

independent due diligence on the deal. In some cases, in an echo of the incestuous web of cross holdings of the 1920s investment trusts, two CDOs often
bought each other's mezz. Often the only 'real money' buyers were those who
bought the senior.

### 8.2.2   Where does the Money come from?

There are more ways to make money in structured finance than simply
taking fees for underwriting ABS issuance. Others included transactions with
securitisation vehicles such as interest rate swaps or liquidity facilities, asset
management fees when – as was increasingly the case – the underlying collateral was managed[307], and secondary market trading.

### 8.2.3   Adverse Selection of Bank Assets

Securitisation typically works best for assets that can easily be understood, and which appear to be low risk.  Thus some of the lowest credit
spreads in structured finance come from the supersenior tranches of the best
credit card deals. Here the underlying assets are highly diverse, short duration,
high yield, and supported by considerable credit enhancement. Moreover investors understand the behaviour of the collateral: there is no need to read
a hundred page book to appreciate its nature. This favourable treatment of
easy-to-analyse, good quality assets means that banks are naturally tempted
to securitise those. The other stuff – assets that are hard to analyse, or bad
quality, or both – are kept on balance sheet. Thus the ODM naturally leads to
a decline in the quality of the assets left on banks' balance sheets[308].

## 8.3   Structured Finance in the Boom Years

In this section we look at how firms used structured finance technology.
The majority of structured finance related losses came from proprietary positions: why did firms have them?  And how did some market participants
avoid the losses that befell the unwary? We look at some of the answers here:
another part of the story is in the next chapter.

### 8.3.1   Delta Hedged ABS Tranches

One good example of delta hedging in structured finance was the AMPS
('amplified mortgage portfolio') positions at UBS[309]. Here the firm had a large
position in the top, AAA-rated tranches of subprime-mortgage-based securities. This was 'hedged' by buying credit protection on the underlying mortgage portfolio as figure 8.3 illustrates.

The firm used a model to predict the loss on the super senior position for a small increase in credit spreads. It then buys enough protection such that the gain in value of the protection is equal to modelled loss.

The position was therefore hedged for small moves in credit spreads. It appears that UBS's risk model therefore assumed that the position was very low risk, and the firm therefore allowed a very substantial position to be accumulated.

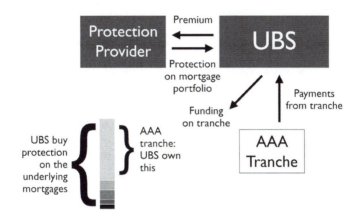

**Figure 8.3**: Schematic of the AMPS trades

The AMPs position was also *positive carry*, meaning that the income from the supersenior position after funding was larger than the cost of the credit protection. This was because the model suggested that each billion dollars of supersenior only required two, three or four million dollars of credit protection as a hedge[310]. Figure 8.4 illustrates the economics of the trade assuming that Libor is 4% and the trade is funded at Libor flat. With a big position – $10B of AAA-rated ABS – the firm would make roughly $16M a year.

Positions like this are perfect for a trader wanting to get a big bonus. It seems to make money every day, yet it does not look risky according to their employer's risk system.

When spreads gapped out, though, (as shown in figure 1.5), the loss on the supersenior was much larger than the gain on the credit protection, and UBS lost a very substantial amount of money. They were not alone in this: when the Crunch came, the combination of large market falls and illiquidity severely challenged the hedging strategy of many market participants. But

| Position | Size | Credit spread | Annual Cashflow |
|---:|:---:|:---:|:---:|
| AAA ABS | $10B | Libor + 12 b.p.s | $420M |
| Funding of ABS | $10B | Libor | ($400M) |
| Protection on pass throughs | $300M | 150 b.p.s. | ($4M) |
| **Net profit per year** | | | $16M |

**Figure 8.4**: A summary of the economics of a delta hedged position in a supersenior tranche if things go well

they were relatively unusual in the sheer size of their portfolio, and hence in the subsequent losses. The use of a single risk measure, based on historically experienced moves in risk factors, seems to have been responsible. It appears that the firm's risk infrastructure simply did not allow for the possibility that the future might not be like the past.

### 8.3.2 Credit Derivatives on the ABX

The AMPS trade brings out a useful feature of credit derivatives from the perspective of a structured finance trader: one can go short – and so profit if prices fall – by buying credit protection[311]. One of the most important tools here are credit derivatives on the various ABX sub-indices. Liquidity was much better here than for derivatives based on individual MBS securities, and so it was possible to buy and sell quite large amounts of protection.

Thus many firms' attempts to hedge their exposure to subprime in 2007 included purchasing protection on the ABX[312]. Since losses hit the lowest tranches first, protection on the tranches rated BBB or BBB- gives a geared exposure to declines in the value of the underlying mortgage pool. Those firms which emerged unscathed from the first wave of the crisis in 2007 were not necessarily those that had little or no subprime activity. Rather it was those that had spotted the rising tide of problems and hedged themselves by buying protection.

Note that the ability to short the ABX is a bit of a mixed blessing. The wave of protection buying meant that the indices fell fast during 2007: perhaps faster than was justified by the decline in value of the underlying securities. Buyers of risk were scarce since most people who might have been interested in MBS already owned it and, lacking good information about the eventual performance of U.S. mortgages, they did not want to increase their exposure even at bargain prices. Thus the ABX tumbled.

This mattered to most holders of subprime risk, not just hedgers. The ABX indices offered one of the few visible prices of mortgage risk, so many market participants used it as the basis of their valuation of subprime MBS. If the ABX fell too far and too fast then those marks would have been too low too.

### 8.3.3  Seizing the Flow

Many firms wanted to capture both the lucrative initial structuring and underwriting fees, and the other opportunities for profit in performing securitisations. All they needed were assets to securitise. By 2006, these were becoming hard to come by. You could not issue a CDO of ABS without ABS to go into it, and investor demand for ABS was fierce. To meet this demand, investment banks went up the origination chain: for instance Merrill Lynch bought First Franklin, a mortgage origination and servicing company in September 2006, primarily to access a bigger flow of mortgages.

Needless to say, these deals went sour as the market fell. Merrill paid $1.3B for First Franklin, and discontinued mortgage origination at the firm less than two years later. In the interim First Franklin provided mortgages for the firm's growing and seemingly profitable structured finance business[313].

### 8.3.4  Structured Finance Risks and their Management

Structured finance techniques allowed the ODM to develop, transforming accrual accounted held-to-maturity risk into multiple shards. Different tranches of risk were held by different players, with a tranche sometimes funded by one party but with the risk taken by another. The use of fair values, both to account for the risk itself; to set margin; and in the determination of events of default[314], meant that when the market fell there was a wave of selling, some of it forced.

Some large financial institutions owned significant notionals of ABS because they could fund it cheaply. Low credit spread volatility and high ratings left an impression of low risk for these positions, and sometimes full benefit was given for hedges of imperfect efficiency or which introduced significant counterparty risk. Then when the Crunch began ABS prices fell and leverage providers saw simultaneous falls in credit quality of unfunded risk takers, rises in margin required, and (to a lesser extent) falls in the value of collateral they had taken.

Firms faced a significant risk management challenge in structured finance. Positions were often complex, highly leveraged, and sometimes only approximately hedged. This gave rise to model risk. These problems were compounded by model to the benign conditions that pertained before the Crunch. Some firms managed these challenges reasonably well, and these institutions tended to share some common features[315]. A report by an international group of supervisors identifies some factors which were associated with institutions who managed these challenges well:

- *Risk culture.* Those firms where risk information was widely shared across the organisation fared better. Firms which managed risk in silos, or where risks were only narrowly discussed, tended to do less well.

- *Valuation risk management.* Firms with sound internal processes for the valuation of complex and/or illiquid securities were more successful. In particular the processes for developing proxies where marks are not available, taking prudent valuation adjustments, and ensuring wide understanding of the uncertainty in marks were important.

- *Funding risk management.* A good integration of Treasury and other liquidity risk management processes into the broader firm risk control process tended to reduce losses.

- *Model risk mitigation.* Successful firms tended to use multiple risk measures, to be sceptical of the veracity of any single risk measure, and to continually refine their risk reporting as new aspects of risk emerged.

These broad factors indicate that it is possible to engage in securitisation and other forms of structured finance safely: it is not the products that are toxic, but rather certain less rigourous management practices.

## 8.3.5    *Size is (nearly) everything*

The subprime-related losses suffered by some firms were huge: some of them were illustrated in figure 6.11. Many of these losses came from positions that firms proactively acquired, such as UBS's AMPs. But there is another factor too: some firms may have kept risk simply because they could not sell it without taking a loss.

Here's how it happens. Suppose the bank is trying to be a leading ABS underwriter. There is considerable prestige attached to a high position in the league tables for underwriting particular classes of debt, and firms strive to be well placed. Unfortunately to be well placed, you have to offer better deals to the people with assets than anyone else. Sometimes, perhaps inadvertently, you will offer an uneconomically good deal. Thus the leading structured finance players inevitably end up with inventory positions which cannot be sold profitably. The more deals you do, the more chance there is of this happening. And when it does, some managers may be tempted to keep the position rather than sell it at a price the market does like and take a loss.

Figure 8.5 shows the top three positions in the league tables for CDOs and for American mortgage backed securities for 2006[316]. While we certainly cannot infer that just because a firm was a leading bookrunner, it had significant amounts of inventory, the high correlation between a firm appearing in the figure and a firm having problems is noticeable.

|   | Worldwide CDOs | U.S. MBS |
|---|---|---|
| 1. | Merrill Lynch | Lehman Brothers |
| 2. | Deutsche Bank | Bear Stearns |
| 3. | Citigroup | RBS |

**Figure 8.5**: Leading bookrunners of structured finance transactions in 2006

In 2007 the leading bookrunners of worldwide CDOs were again Citigroup and Merrill Lynch, each being responsible for almost $40B of ABS. Perhaps this helps to explain their position in the table of write downs (table 6.11).

## 8.4    Insurance in Form and Name

An insurance company does three things.

1.    It takes risk by *writing insurance*;

2.    It *invests the premium* received;

3.    And it *pays claims* when they arise.

### 8.4.1    Taking Insurance Risk

The process of deciding the appropriate premium to charge for an insurance policy is known as *insurance underwriting*. Typically here an *actuary* will perform a historical analysis of the likelihood and timing of claims on the particular type of insurance contemplated. Thus for a life insurance policy the likelihood of a claim at some point is 100%, but the timing varies[317].

Most people are familiar with some types of insurance: life, property, and catastrophe insurance immediately come to mind. Insurance companies also sometimes write various forms of commercial insurance including maritime and liability insurance[318]. In all of these cases the past is a reasonable guide to the future. There may be slow trends, such as people tending to live longer as health care improves, but broadly the behaviour of a large collection of life insurance policies written in the U.S. in 2008 is likely to be similar to those written a few years earlier or later.

It is important for an insurance company to have a large, diverse portfolio of exposures. Writing one policy which pays out a billion dollars if a Richter force 6 or above earthquake hits central LA is very risky. The insurer may well fail if the San Andreas fault shifts. It is better to write hundreds of US, Japanese and South East Asian earthquake polices along with European windstorm, hurricane cover, satellite, and hailstorm cover, none with a payout of more than twenty million dollars. This might well be a diversified portfolio of catastrophe risk. Diversification allowed actuaries to be reasonably confident that their firm was being properly compensated for the risks it was taking and that it was unlikely to fail[319]. Moreover, if one risk assessment is wrong, a diversified portfolio is safer since the different risks are fairly uncorrelated. Getting the price of Florida hurricane protection wrong does not affect your LA earthquake position.

### 8.4.2  Traditional Insurance

Insurance can be a profitable business: some insurers are very large, valuable companies. The insurance business is usually divided into two types: life insurance and general insurance. The success of both types of insurer can be seen as validating the approaches of both life and general insurance actuaries.

### 8.4.3  Insurance Accounting

Insurers recognise profit and loss in a rather different way from securities or derivatives traders. The details vary from country to country, and are anyway in a state of flux[320], but the key issues are:

- Insurance contracts are generally not marked to market; instead, premium income is recognised as it is earned; and

- Provisions are taken for claims only as they become highly likely[321].

If you buy a bond issued by a company, as that company slides towards bankruptcy, you recognise losses as the credit spread of the company increases and the bond falls in value. If you write an insurance policy which protects against loss on that bond, you may well not recognise any losses at all until and unless there is a claim on the policy[322].

### 8.4.4  Credit Derivatives and Insurance: Similarities and Differences

The similarities between a CDS on a bond and a financial guarantee insurance policy on the same bond are considerable: both the insurance contract and purchased credit protection pay out if a bad thing happens. These similarities led some insurance companies to think that they had the skills to trade credit derivatives.

The differences between these two instruments are however significant: it is only possible to purchase insurance on a risk that you have, so that you can insure a bond that you own against the risk of loss due to a credit event, but you cannot benefit from insurance on a bond that you do not own. A credit derivative in contrast returns the same amount whether you have bought protection on an asset you own or simply bet that a credit event would happen. Moreover insurance companies typically demand proof of loss, whereas there is no such obligation in a credit derivative[323].

It is rather important for the market that credit derivatives are not insurance, as to sell insurance you need to be an insurance company. Many protection sellers are banks or hedge funds, and should credit derivatives be characterised as insurance, these contracts might not be enforceable[324].

## 8.5   The Rescue of AIG

The American Insurance Group, known as AIG, was one of the largest insurance companies in the world. It was active in nearly every area of insurance including life, commercial and disaster insurance, and it had over 70 million clients. In 2005 it had a AAA rating: something that was a lot harder for a company to acquire than for a securitisation tranche. AIG was also one of the most profitable insurers in the world. Yet by late 2008 it had been forced to sell 80% of itself to the U.S. government in exchange for an emergency loan. How did a company with over a trillion dollars in assets fail so dramatically, and why did the FED save it?

### 8.5.1   AIG before the storm

In addition to its two major lines of traditional insurance business, general and life, AIG had two other large business segments:

1.   *Asset management* involved both investing AIG's premium income to increase its claims paying ability, and investing funds on behalf of third parties.

2.   The *financial services business* was composed of aircraft leasing, consumer finance, and the ultimate source of AIG's problems, AIG FP. FP, for 'Financial Products', was an active participant in the capital markets, trading both securities and derivatives.

### 8.5.2   Financial Products

There were a number of parts to AIG FP's activities, but the key one was writing CDS protection. The group entered into over $500B of CDS[325]. AIG FP was therefore a large CDPC sitting inside the larger AIG. Since the insurer was very highly rated, AIG FP was a sought-after counterparty: the market considered that there was little counterparty risk in dealing with a AAA rated company. FP, then, was a significantly more successful business as part of AIG than it would have been on a standalone basis.

Unfortunately, the skills necessary to run and control a CDPC are significantly different from those necessary to be successful in general or life insurance. Credit derivatives are fair value instruments: they must either be marked to a market price, if one can be found, or else to a price obtained from a model which is calibrated to the market. This practice is very different from the insurance accounting used for AIG's traditional business. Therefore it is perhaps not wholly a surprise that AIG consistently failed to correctly account for its derivatives activities[326]. Material weaknesses were found in the firm's internal controls around these parts of the business, and in mid 2005 AIG was forced to restate its earnings for each of the years from 2000 to 2004, reducing its income by $3.9B.

Understandably, this was not taken well. AIG lost its AAA rating, in part due to this news.

### 8.5.3   The Danger of Downgrades

Many counterparties were willing to deal with a AAA-rated AIG on an uncollateralised basis. But some anticipated the risk that AIG's credit quality might decline. Therefore they built a proviso into their credit support annexes with AIG: the insurer would not have to post collateral as long as it was highly rated, but it would have to put up money against its derivatives if its rating fell too far. This is a particularly nasty form of liquidity risk: just when a firm is finding it harder to fund itself, thanks to the downgrade, it has to find extra funds to meet these *downgrade triggers*. As the firm's 2007 annual report put it, 'AIG's liquidity may be adversely affected by requirements to post collateral.'

### 8.5.4  Shares Slide, System Stresses

AIG FP's activities included writing CDS on the supersenior tranches of CDOs. Many of these CDOs had subprime MBS in their collateral pool. The prices of the underlying ABS bonds fell during the Crunch, so the CDS AIG FP had written became more valuable to the holder. AIG FP therefore suffered mark to market losses: $11B in 2007, and more than $5B in the first half of 2008[327]. AIG also announced in February 2008 that it had found further accounting weaknesses, and the firm took another earnings adjustment of nearly $5B[328].

If there is one thing worse than restating earnings due to accounting errors for a company's credibility, it is repeatedly restating earnings. AIG was downgraded again, and the stock price fell significantly.

The downgrades caused AIG to have to post significant collateral. Moreover its announced losses had alerted the market to the extent of the subprime risk in AIG FP's portfolio of written default swaps, and hence to the possibility of further losses and further downgrades. As liquidity became harder to find and more expensive in late 2008, AIG's share price fell fast. The vicious circle was completed by an additional downgrade: on 15th September 2008 the insurer was downgraded again partly due to its 'extremely limited' ability to raise cash and its declining stock price.

Thus a falling stock price, due to fear of a downgrade, meant that the firm could not easily raise money, which meant that it was actually downgraded, forcing it to post more collateral: over $10B more. This made its funding problem far worse and depressed the share price even further. Figure 8.6 illustrates the share price and the timing of the agency downgrades.

### 8.5.5  $85B at Credit Card Rates

This last downgrade brought AIG close to failure. It could not raise the extra sums needed to meet collateral calls easily, and a credit support default was entirely possible. This would have caused significant systemic risk: AIG FP had written far too many default swaps for its failure to be easily absorbed by the financial system. Counterparties who had been relying on AIG to make good their losses would suddenly find themselves unprotected. Fear of the consequences of this led the FED to intervene. On the 16th September, the Federal Reserve Bank of New York allowed AIG to borrow up to $85B against collateral[329]. The FED facility was for two years, and AIG had to pay interest on it at Libor plus 850 basis points. This enormous spread reflected the usual lender of last resort doctrine that emergency funding should be at a penal rate.

**Figure 8.6**: The AIG share price and downgrades of the firm's credit rating

The U.S. government did not rescue the shareholders of AIG: to do so would have introduced far too much moral hazard[330]. In exchange for the loan, the government received a 79.9% equity interest in AIG and the right to veto the payment of dividends to common and preferred shareholders.

### 8.5.6  Aftermath

AIG certainly needed the FED's money: two weeks after the bail out, it had already drawn down $61B of the permitted $85B, and by the 8th October it was forced to ask for the loan of an additional $37.8B. The firm placed many of its units up for sale, undertaking only to retain its general insurance business. Everything else, including AIG FP, was either up for sale or being wound down. As the new chief executive said, though, 'It is not something you can announce on Friday and expect to do by Monday'[331].

What did AIG do wrong? Clearly the firm did not understand the risk of a liquidity/downgrade/share price spiral. Without collateral on downgrade clauses in its CDS, it might not have needed the bailout. Moreover the sheer size of its CDPC was enormous even if the risks in it were diverse. And failing to fix the firm's internal controls after the 2005 earnings restatements was clearly unfortunate. Given this, it seems bizarre that the ex-CEO of the firm blamed not himself and his management team, but accounting rules. Speaking in front of the Waxman Committee hearings[332], Martin Sullivan seemed upset that AIG was 'forced to mark our [credit default] swap positions at fire-sale prices as if we owned the underlying bonds, even though we believed that our swap positions had value if held to maturity'. Perhaps in future firms may consider that an understanding of the dynamics of profit and loss, and how unrealised losses affect confidence and their rating, are just as important as their beliefs about eventual asset value.

## 8.6   Off Balance Sheet Funding

Liquidity is useful. The conventional ways of getting it, like borrowing from a bank or issuing a bond, are *on balance sheet*. That is, they appear as a liability of the firm and contribute towards investors' assessment of the firm's gearing. Wouldn't it be nice to be able to fund assets and get the benefit from them without the associated borrowing being so obvious? Off balance sheet funding techniques achieve this.

### 8.6.1   ABCP

An important tool in this area is *asset backed commercial paper* or ABCP. Like ordinary commercial paper, this is a short term debt instrument issued to investors and regularly rolled. Unlike standard CP, it is backed not by the promise to repay of a firm, but by specific assets. Thus ABCP is short-dated debt issued by a securitisation SPV.

The main difference between a typical securitisation and an off balance sheet funding using ABCP is maturity mismatch. A securitisation takes assets with a certain duration – five years say – and issues bonds with the same maturity to fund them. The securitisation SPV is in that sense a zombie: once it has bought the collateral and issued the debt, there is nothing for it to do except drip payments from the collateral down the waterfall and make the resulting tranche payments. In an ABCP deal, in contrast, the liabilities expire regularly, so new ABCP has to be issued to refinance the vehicle. These zombies need to feed regularly.

### 8.6.2   Conduits as Old Style Banks

A securitisation SPV which issues ABCP is known as a *conduit*. The simplest form involves a sponsoring bank selling collateral into an SPV. Typically the collateral is diverse, so for instance the bank might transfer some credit card receivables, some loans, and some mortgages. The vehicle issues ABCP to fund 90% or more of the purchases, the rest coming from its equity (which is mostly kept by the bank[333]).

The economics of this transaction are that payments from the collateral first go to pay interest on the ABCP, with any excess above this belonging to the equity holder. The ABCP pays interest based on short term rates, whereas the collateral's return is based on longer term rates. The SPV's net interest income is comprised of a term structure component and a credit spread component, just like the old-style bank's net interest income we discussed in section 4.1.6. In fact a conduit is an old style bank in miniature: figure 8.7 illustrates the key similarities.

|  | Old style Bank | Conduit |
|---|---|---|
| Equity Capital | Provided by shareholders | Provided by sponsor |
| Debt | Short term deposits | Short term ABCP |
| Assets | Long term loans | Various long term ABS |
| Income | Term structure pickup in funding plus credit spread income | |

**Figure 8.7**: Conduits as off balance sheet old style banking

In the good times, conduits and related structures allowed banks to enhance their earnings by benefiting from the return of the conduit's assets but without having to hold those assets themselves. This meant in particular that the banks could avoid regulatory capital requirements on those assets. It seemed that conduits offered all of the benefits of old style banking with none of the troublesome restrictions.

### 8.6.3    SIVs

A *structured investment vehicle* or SIV is similar to a conduit, but rather than just having two elements in its capital structure – equity and ABCP – it has three or more. There are two ideas here.

1.    By introducing a mezzanine layer of capital, the senior debt can be highly rated despite there being only a small amount of equity. This dramatically enhances leverage: over one hundred to one is possible[334].

2.    The mezzanine notes are typically longer duration. They provide a higher yielding, long term instrument for investors who want something riskier than the ABCP.

## 8.6.4 What the Vehicles bought

Conduits and SIVs had a range of investments. Most conduits had a significant amount of trade receivables, while SIVs often bought financial institution debt as well as ABS. However they both invested in MBS, and in structured finance paper. Figure 8.8 gives some estimates[335].

|  | SIV Assets | Conduit Assets |
|---|---|---|
| Mortgage risk | 36% | 8% |
| CDO Tranches | 15% | – |
| ABS | 12% | – |

**Figure 8.8**: Estimates of SIV and Conduit investments in structured finance instruments for 2007

This collateral composition meant that most SIVs suffered significant losses as MBS, CDO tranches and other ABS fell in value during 2007. Conduits were less exposed, but a fifty percent fall in 8% of assets is still a 4% fall in value, enough in many cases to blow through the conduit's equity. These vehicles were large too: $10B or more of assets was perfectly common.

## 8.6.5 SIV and Conduit Risks

SIVs and conduits have a nasty type of funding liquidity risk. Unlike old style banks, which have retail deposits with government deposit protection, the majority of the funding of both SIVs and conduits is both short term and highly confidence sensitive. SIVs and conduits need to roll their ABCP frequently, often every three months. Their ability to do that depends on their credit quality: ABCP investors want to be convinced that the vehicle's assets are worth significantly more than the ABCP. Therefore if asset values dip even a little – and given that there is only a slim layer of equity – it can suddenly be impossible to roll funding. An ABCP buyers' strike is therefore the analogue of a run on an old-style bank.

The legal structure of some SIVs and conduits also leads to a new risk. Some of these vehicles were required to prove regularly that they had a sufficient degree of over collateralisation. That is, they had to mark their assets to market, and show that the value of these assets exceeds, say, 102% of the value of their liabilities. The problem with this was that as the Crunch hit, establishing asset values became difficult. The collateral became illiquid, and the vehicle's solvency (let alone whether they had enough over-collateralisation) became a matter of debate.

Finally in some cases SIVs and conduits were used as dumping grounds. That is, when a bank's structured finance group produced a security that could not be sold – either because the market thought it was too expensive or the structure was unattractive – then the paper was put into the bank's SIV[336]. Thus there is a suspicion, at least, that some SIVs and conduits held not a diversified portfolio of structured finance risk, but a portfolio of structured finance risk that no one else wanted to buy.

### 8.6.6   ABCP rolls and Backup Liquidity

Suppose that an off balance sheet vehicle has good quality assets but ABCP buyers, in an act of caprice, refuse to allow it to roll its paper. The vehicle will then be forced to liquidate its assets. Forced liquidation often does not achieve the full value available in an asset portfolio, so many vehicles were structured with *backup liquidity lines*. This was a facility whereby a conduit could borrow from one or more banks if it could not roll its ABCP.

Concern rose among ABCP investors during 2007. They increasingly withdrew from the market. This triggered backup liquidity lines. Often (but not always) these lines had been provided by the sponsoring bank. As soon as a sponsor had a significant equity stake in a vehicle and most or all of the vehicle's funding, then they effectively owned the vehicle's assets. Off balance sheet had become on balance sheet again and the banks ended up owning assets just when they were falling in value and becoming less liquid. An expert report puts it well:[337]

> It is widely recognized that a major contributing factor to
> the credit market crisis was the manner and extent to which
> risks associated with certain classes of off-balance sheet ac-
> tivities at many financial institutions may not have been
> adequately encompassed within firms' risk monitoring and
> risk mitigation frameworks. Thus, when contingent risks
> arising from such off-balance sheet activities were triggered,
> this gave rise to elevated contagion risk and large write-
> downs at many financial institutions.

### 8.6.7   Stand or Delever?

Assets did not come back on balance sheet solely as a result of backup liquidity lines. Even without them, the vehicle sponsors faced problems. On the one hand, they were only legally exposed to whatever equity they had retained. If there were sufficient losses to blow through the equity, then the ABCP buyers would suffer. But the sponsors used the CP markets themselves, and ABCP had sometimes been sold as a low risk investment. Could the banks really leave ABCP holders with substantial losses?

Most banks judged that they could not. The reputational impact would be too great, and the threat to the bank's access to the CP market was too troubling. So the banks bought back the ABCP at par, and took conduit and SIV assets back onto their balance sheets, suffering losses in the process.

This exposed the farrago of off balance sheet funding. The banks had told their regulators and auditors that the ABCP buyers were really taking risk, the assets had really been sold, and so there was really no need to take regulatory capital against the assets nor to count the ABCP as a bank liability. But when it turned out that the ABCP investors might actually lose something, liquidity lines were tripped and/or the banks (mostly) provided support to their vehicles, ensuring that the investors did not suffer[338]. It seems that no risk transfer had actually taken place.

### 8.6.8 ABCP Fades Away

The unwinding of the off balance sheet funding vehicles can be seen in the amount of ABCP in issuance. By mid 2007 there was $1.2T of ABCP in issuance[339]. ABCP grew from 39% of the total commercial paper market in 2000 to move than 55% in mid 2007. By the end of 2008, though, the total volume had fallen to $700B, 45% of the total[340]. The importance of off balance sheet funding for banks was declining.

Bank investors had reason to be thankful for this. Banks' disclosure on their involvement with off balance sheet vehicles has typically been scant, and it was very difficult for investors to assess the extra risks these entanglements generated. For instance Citigroup was the sponsor of a number of large conduits and SIVs, yet their 2006 annual report – written before problems arose in the ABCP market – simply devotes a few lines to these relationships with little risk disclosure beyond the total notionals involved.

### 8.6.9 Where did subprime risk go?

We now have all the ingredients to understand who ended up holding subprime risk.

- Some of it was kept by the originator. But much of it was passed on and securitised.

- The vast majority of the securitisations had some form of credit enhancement. This was often provided by the originator or by the aggregator, so these parties ended up with significant amounts of risk.

- The MBS from securitisations was bought by end investors, hedge funds, government sector entities (including Fannie and Freddie) and structured finance players.

- This last class of buyer used the MBS as collateral in CDOs of ABS, in negative basis trades and other forms of unfunded risk taking, and as part of the collateral pool for off balance sheet vehicles. On occasion they kept risk that could not be sold or that they decided offered an attractive risk/return trade-off.

### 8.6.10   The MLEC, R.I.P.

Conduits and SIVs started to have problems rolling ABCP in early 2007, and by autumn of that year, the problems were widespread. Many large banks had SIVs, conduits, or both, and there was concern that if all of these vehicles were liquidated at once, the result would be a flood of ABS entering the market at once and further depressing prices.

At this point the U.S. Treasury stepped in, hosting discussions with the most affected banks. The result of these meetings was a plan to set up a large conduit to buy ABS. This *Master Liquidity Enhancement Conduit* or MLEC would buy good (meaning reasonably highly rated) assets from distressed SIVs and conduits and thus prevent the wave of liquidations. Three banks, Citigroup, JPMorgan and Bank of America, supported the plan. It was hoped that the MLEC might increase confidence sufficiently that ABS (and ABCP) market liquidity would improve.

There were a number of problems. First, the MLEC was supposed to be self funding. Why would investors who were reluctant to buy ABCP from ordinary conduits be comfortable with the assets in the MLEC? This was especially an issue as the MLEC would naturally attract the least liquid and hardest to value ABS. Second, given that the MLEC was supposed to be a buyer of last resort, how would it establish the right price for these illiquid securities? If the price the MLEC was willing to pay was too low then no one would sell to it: but if it was too high, then the MLEC would require massive amounts of equity to absorb its losses.

These problems proved fatal. No European banks joined the plan, and soon nothing more was heard of it. Individual banks dealt with their own vehicles, and the ABS market struggled on. It remained illiquid, but – to the extent that any trading happened – prices did not crash much further than they already had.

## Was Structured Finance The Culprit?

No. A product by itself can never be responsible for a problem in the financial markets since they do not trade themselves. Rather it is the incentives that encouraged banks to move risk to insurance companies, ABCP investors, and other investors that caused the problems.

Structured finance is a tool. It can make credit cheaper by allocating it more efficiently. But it can also create new risks which are not always understood by all parties, such as credit support default risk, or the risk of backup liquidity lines. Structured finance also created highly complex structures and trading strategies which introduced model risk. The valuation and risk assessment of these position often depended on liquidity assumptions which some firms did not understand. Thus while structured finance itself is not bad, it was certainly possible to use it badly.

## Notes

[287] For a complete list and a more extensive discussion, see section 2.5 of my book *Understanding Risk*.

[288] This glosses over the details a little. In fact typically any obligation that ranks *pari passu* with the reference obligation is deliverable. The reference obligation is used simply to determine if there has been a credit event.

[289] It can cause a loss if we have bought the bond at a premium: if we buy the bond today for 101 and it defaults tomorrow, we deliver it and receive the face value of 100, so the 1 of premium is lost.

[290] In particular funding costs and the difficulty of shorting corporate bonds mean that the pure credit spread of the bond – the par asset swap spread – can differ from the CDS spread. Furthermore while nearly all events which could affect a bond holder's ability to recover their principal are covered by CDS credit events, there are also situations – such as some technical restructurings – which constitute a credit event but do not substantially affect the bond price. See Paul Harding's *A Practical Guide to the 2003 ISDA Credit Derivatives Definitions* (2004) for more details.

[291] The situation is actually rather complex since the protection would probably have been purchased from one of the operating subsidiaries of Lehman Brothers Holdings, Inc. rather than the parent company. The parent entered into a U.S. chapter 11 filing, but some of the operating subsidiaries did not. In particular Lehman Brothers International (Europe), one of the main Lehman derivatives traders, is being liquidated under UK law. Typically derivatives claims rank *pari passu* with senior debt and default is an early termination event, so a party who had bought CDS protection would have a senior claim equal to the mark

to market of the CDS. The claim would be against the entity it had traded with, not on the parent Lehman entity. For more details, see Daniel Passage and Francisco Flores' *Financial Agreements with Lehman Brothers* (O'Melveny & Myers 2008) available at `www.omm.com/financial-agreements-with-lehman-brothers-09-18-2008` or Richard Rosenstein et al.'s *Effect Of Lehman Brothers Holdings Inc. Bankruptcy On Swap Agreements* (Nutter, McClennan and Fish, 2008) available via `www.nutter.com`.

[292] See the International Swap Dealers Association Press Release *ISDA Mid-Year 2008 Market Survey Shows Credit Derivatives at $54.6 Trillion* (September 2008) available at `www.isda.org/press/press092508.html`.

[293] For a further discussion, see Michael Gibson's *Credit Derivatives and Risk Management*, (Federal Reserve Board Finance and Economics Discussion Series 2007) available at `www.federalreserve.gov/pubs/feds/2007/200747/200747pap.pdf` or Frank Partnoy and David Skeel's *The Promise and Perils of Credit Derivatives* available at `papers.ssrn.com/sol3/papers.cfm?abstract_id=929747`.

[294] This is not strictly true as any CDS payouts would have to be funded.

[295] See the *User's Guide to the 1994 ISDA Credit Support Annex*, International Swap Dealers Association.

[296] For more on the risk of CDPCs, see *Criteria For Rating Global Credit Derivative Product Companies* by Standard and Poor's, available at `www2.standardandpoors.com/spf/pdf/fixedincome/CriteriaForRatingGlobalCreditDerivativeProduct.pdf`.

[297] See *Credit Derivatives: The Definitive Guide* by Jon Gregory (2003) for more information.

[298] We could use interest rate swaps to move from fixed to floating, but there may not be sufficient cashflow from the collateral to pay the swap coupon: thus we have credit contingent interest rate risk.

[299] For further information on synthetic CDOs and other aspects of credit derivatives, see Gregory or, for a more historical perspective, *Credit Derivatives: An Overview* by David Mengle, available at `www.frbatlanta.org/news/conferen/07fmc/07FMC_mengle.pdf`.

[300] The SPV pledges the cash obtained from the sale of the tranched ABS to the asset originator as collateral against the portfolio default swap

[301] See chapter 2 of my book *Understanding Risk* for a fuller description of hedging.

[302] Such as a Value at Risk model calibrated to ABS data from 2003-2006.

[303] Sometimes the level one ABS are pass throughs rather than tranched securities, so there is not necessarily tranching layered on tranching. But there often is.

[304] For an excellent account of the rôle of subprime mortgages in structured finance, see *Understanding the Securitization of Subprime Mortgage Credit* by Adam Ashcraft and Til Schuermann, Federal Reserve Bank of New York Staff Report No. 318 (March 2008) available via `www.newyorkfed.org/research/staff_reports/sr318.html`.

[305] According to an Intex query run in October 2008, the total direct subprime and Alt-A in CDOs was 67%. However this understates the issue as some of those CDOs will in addition contain tranches of other CDOs which in turn contain non-prime MBS.

[306] If the collateral was in any way managed – which it often was in CDO of ABS transactions – these fees could be structured as charges for asset management. Note that if the manager owns the junior, then their incentive is to select high spread collateral (to keep the excess spread they receive high) rather than safe collateral.

[307] That is, rather than being static, an asset manager may be appointed to trade the collateral pool. These managers are usually either the underwriting investment bank or parties friendly to the bank.

[308] For a further discussion, see Antje Berndt and Anurag Gupta's *Moral Hazard and Adverse Selection in the Originate-to-Distribute Model of Bank Credit* (2008) available at `papers.ssrn.com/sol3/papers.cfm?abstract_id=1290312`.

[309] See *Shareholder Report on UBS's Write-Downs* (2008) available via `www.ubs.com` for more details. Similar considerations apply to correlation positions such as long supersenior 'hedged' with a short junior tranche position.

[310] As the shareholder report says, 'This level of hedging was based on statistical analyses of historical price movements that indicated that such protection was sufficient to protect UBS from any losses'.

[311] This is particularly useful as one cannot typically short ABS in the same way as one can short equity: there is little or no borrow market.

[312] See for instance the New York Times Dealbook article *Banking on the Credit Meltdown*, 9th August 2007, available at `dealbook.blogs.nytimes.com/2007/08/09/banking-on-the-credit-meltdown`.

[313] Announcements of both the acquisition and the subsequent closure of origination are available via `www.ml.com`.

[314] Some SPVs were set up so that if the fair value of their assets fell beneath a fixed value based on their liabilities, they were in default.

[315] This list is condensed from a longer discussion in the Senior Supervisors Group's *Observations on Risk Management Practices during the Recent Market Turbulence* (March 2008), available via `www.fsa.gov.uk`.

[316] The data is from Asset Backed Alert: see `www.abalert.com/Public/MarketPlace/Ranking/index.cfm`.

[317] Thus life insurance can be seen as a race between invested premium, which (if it is invested wisely and the markets are kind) grows every year, and the claim. The earlier the claim occurs, the less time the invested premium has had to grow, and thus the more likely a loss is to the insurance company. Life insurers therefore have *longevity risk*: shorter lives cost them money. See Heather Booth et al.'s *Lee-Carter mortality forecasting: a multi-country comparison of variants and extensions* (Demographic Research 2006) available at `www.demographic-research.org/Volumes/Vol15/9/15-9.pdf` for a discussion of longevity forecasting.

[318] Liability insurance protects the insured against losses due to some form of liability. Examples include product liability (where the loss may for instance be due to failure to adhere to consumer protection legislation), employment liability, or director's liability insurance.

[319] In fact, on average, insurance companies charge less than the fair value of claims. What saves them from insolvency is that invested premium increases in value sufficiently to cover the difference and to make them some money.

[320] For a summary of the state of play on the current extended efforts to improve insurance accounting, see `www.iasb.org/Current+Projects/IASB+Projects/Insurance+Contracts/Insurance+Contracts.htm`.

[321] It should be noted, though, that provisions play a proportionately larger role relative to capital for insurer than for banks: see *Risk Management Practices and Regulatory Capital – Cross Sectoral Comparison*, Basel Committee on Banking Supervision, 2001 available at `www.bis.org/publ/joint04.pdf`.

[322] You may well think that this is odd, and I agree. Certainly Tom Jones, vice chairman of the International Accounting Standards Board, thought that 'Insurance accounting around the world is broken'. See uk.reuters.com/article/bankingFinancial/idUKN2932543520 080429.

[323] The Financial Services Authority's Discussion Paper 11, *Cross-sector risk transfers* (May 2002) available at www.fsa.gov.uk/pubs/discussion/dp11.pdf gives a further discussion of the relationship between credit derivatives and insurance. The difference between the legal frameworks for insurance and derivatives, whereby an insurance purchaser may have the duty of utmost good faith to disclose facts material to the insurance seller, but no such duty is incumbent on the credit derivatives buyer, is also significant.

[324] Specifically protection buyers may be able, at their option, to either enforce claims or demand that premiums are refunded. The decision by the New York State Insurance Department that from 1st January 2009 they would treat purchased CDS protection where the protection buyer owns the reference instrument as insurance therefore poses a considerable obstacle to the market. Insurance is typically a matter of state rather than federal law in the U.S., so it is entirely possible that the banks could simply move their credit derivatives trading desks across the Hudson River to avoid this snag. A Federal response which mitigates counterparty risk as well as improving regulation – such as setting up a clearing house for credit derivatives – would seem to be a better solution. See *Revisited: Credit Default Swaps - Are They Contracts of Insurance?* from Mayer Brown JSM, available at www.mayerbrown.com/publications/ article.asp?id=5694&nid=6 for a discussion of the legal issue.

[325] $561B according to the 2007 AIG annual report. Not all of these positions would have been written protection, but many of them were.

[326] See *AIG announced delay of third quarter form 10-Q, Restatement relating primarily to derivatives and hedged items and estimates of third quarter results*, AIG press release, November 2005, available via www.aig.com.

[327] See AIG's 2007 Annual Report and second quarter earnings available via www.aig.com for a comprehensive discussion.

[328] According to CNN, the precise issue in early 2008 was whether CDS protection written on a bond should be valued using the CDS spread or the bond spread: see money.cnn.com/ 2008/02/11/news/companies/boyd_aig.fortune/index.htm.

[329] See the FED's press release available at www.federalreserve.gov/newsevents/press /other/20080916a.htm for more details.

[330] Of course any rescue involves some moral hazard. Nevertheless, by ensuring that shareholders are massively diluted, this is reduced. The FED had earlier refused a loan to AIG: it was only when the risk due to AIG's connections with the rest of the financial system became clear that the FED lent.

[331] See the transcript of the 3rd October 2008 AIG Business Update Call, available at seeking alpha.com/article/98457-american-international-group-business-update-call-transcript.

[332] See the Waxman Committee hearing on the Causes and Effects of the AIG Bailout, further details of which can be found at oversight.house.gov/story.asp?ID=2211.

[333] Again, accounting rules mean that things are not quite so simple as we have described: in order to get deconsolidation, the bank cannot keep 100% of the equity. The solution is either to sell some of the equity to third parties, or to band together with some partners to produce a multi-seller conduit.

[334]Douglas Long, in *Converging Developments in the ABCP Conduits and SIV markets* (2006) quotes the leverage of Sedna, a Citigroup SIV, as 149:1.

[335]The breakdown for SIVs is from Box 2.5 of the IMF's *Global Financial Stability Report* (April 2008). Other data suggests that the IMF figures for conduit investments in structured finance paper are low.

[336]This explains the percentage of trade receivables in conduits. This is short maturity collateral which requires a lot of work to analyse properly. Often diversification in trade receivables portfolios is poor due to systematic risks. This makes trade receivable backed tranches unattractive for many investors.

[337]See the Report of the CRMPG III, *Containing Systemic Risk: The Road to Reform* (2008) available at www.crmpolicygroup.org/docs/CRMPG-III.pdf.

[338]Support to a securitisation SPV or ABCP conduit that is provided voluntarily by the sponsor (that is, without a legal obligation to do so) is known as *implicit support*.

[339]This data is from Federal Reserve reporting on Commercial Paper Outstanding available at www.federalreserve.gov/releases/cp/outstandings.htm.

[340]The decline of ABCP would probably be even more pronounced without the FED's Asset-Backed Commercial Paper Money Market Mutual Fund Liquidity Facility. This allows banks to borrow from the FED using ABCP as collateral, and hence it encourages them to provide liquidity to the ABCP market. See www.federalreserve.gov/monetarypolicy/abcpmmmf.htm for more details.

# Chapter 9

---

## Municipal Finance and The Monolines

> *A writer must be as objective as a chemist: and*
> *he must abandon the subjective line;*
> *he must know that dung-heaps play a very reasonable part in a landscape,*
> *that evil passions are as inherent in life as good ones.*
> Anton Chekhov

---

## Introduction

We have already seen how involvement in structured finance brought AIG to the edge of failure. Here we look at another disastrous entanglement: the activities of certain insurers – the *monolines* – in structured finance. These insurers used to operate in a relatively quiet backwater of the capital markets, *municipal finance*. This chapter discusses the origins of municipal finance, the growth of the monolines, their diversification into structured finance, and its consequences.

A number of people foresaw trouble in the growth of the monolines' portfolios of ABS risk. But not many of them made a billion dollars from their foresight. We look at one of the ones who did. This provides a good example of the kind of clear sighted risk analysis that was far too uncommon in the Boom years.

The monolines provided credit enhancement for a particular type of bond used in municipal finance, the *auction rate security*. When concerns started to arise about the health of the monolines, the market for these bonds suffered considerable disruption. The flaws in the design of auction rate securities that made this possible are discussed.

Finally, the travails of the structured finance market from this chapter and the last are pulled together as we look at why this system of finance was so vulnerable to a Crunch.

## 9.1   Municipal Finance

Suppose that a town needs a new bridge. The town council decides that it can afford the bridge. A contractor who can build it is identified. The project will probably be successful – bridge building is reasonably well understood – but there are risks. In particular if the contractor goes bust before the bridge is built, then the town may have to pay someone else to finish it. Perhaps the council should not be allowed to risk municipal funds in this way. But what is the alternative?

### 9.1.1   Sureties and the Monolines

A *surety bond* is a promise by a third party to fulfil a contract if someone else doesn't. The surety takes a fee in exchange for making this promise. Thus a surety bond on a bridge building project would be a promise that if the contractor does not complete the bridge to the agreed specification, then the surety will resolve the issue, leaving the town with an acceptable bridge.

A surety bond is very much like an insurance contract: a premium is paid in order to pass on the risk of a fairly unlikely but possibly catastrophic event to a third party. Thus municipal taxpayers are spared the risk of having to pay for their bridge twice – once to the original contractor, once to a second company who steps in when the first one fails – but they have to pay a fee for this risk reduction.

Sureties are required for many American federal and municipal construction projects. The Federal Miller Act of 1935[341] required that sureties be in place for all Federal civilian construction projects over a certain value, and various states passed mini-Miller Acts imposing the same requirements for municipal endeavours.

### 9.1.2   Financial Guarantees

Surety bonds have been familiar to American insurers for many years thanks to the Miller Acts. Surety writers had acquired considerable competence in assessing the risk of non-performance in a wide range of municipal settings. Thus when municipalities started issuing bonds to pay for their projects, the surety writers were natural homes for the risk of these bonds.

A form of surety bond known as a *financial guarantee* plays a key rôle here. This is a promise to make good any failure to perform on a bond. Suppose the town of Riverville issues a bond to finance the construction of its bridge. Since the bridge is not large, the sum borrowed might be small in bond market terms, perhaps $25M. This makes the Riverville bond problematic: for many investors it is not worth going to the trouble of analysing an issue that is that small. Riverville will suffer from a *coordination problem*: it will have to pay quite a lot to borrow, despite being good credit quality, as no single investor has the resources to determine that the debt is indeed good.

The usual solution to this problem is to pay one party to do the analysis for everyone[342]. In Riverville's case the town identifies a surety who will write a financial guarantee on its bonds. That surety is well known in the bond markets and highly rated, typically AAA. Riverville pays the surety a fee in exchange for an unconditional and irrevocable guarantee that its bonds will pay timely interest and ultimate principal. These guaranteed or *wrapped* bonds then trade based on the credit quality of the surety, that is as AAA-rated instruments.

This reduces costs for everyone. Riverville pays less for the surety than the credit spread it would have to pay without the financial guarantee; the liquidity of the wrapped bonds are good because many investors are happy with the credit of the surety; and the sureties develop considerable expertise in the analysis of municipal finance risk.

Notice that in a financial guarantee, the surety stands in the place of the bond issuer: they only pay as and when the original payments become due. Thus if Riverville issued a ten year bond to pay for its bridge and defaulted three years later, the surety would be obliged to continue paying interest on the bond for the remaining seven years, with principal being due only at maturity.

Note too that a financial guarantee is legally insurance since it is packaged up with the thing it is protecting.

The timing of payments on financial guarantees gives them rather better liquidity characteristics for the writer than those of a credit derivative. In a CDS, the writer has to find the money to make a payment soon after the credit event. In a financial guarantee, large payments are typically only made years later, giving the writer ample time to find the money. This in turn means that there is more credit risk in a financial guarantee: after a failure to perform, the insured party is exposed to the credit risk of the surety for an extended period.

### 9.1.3 The Monoline Insurers

A group of companies grew up to meet the demand for financial guarantees on municipal bonds. These are known as *monoline* insurance companies or just monolines, as they do only one line of business. This is mostly for regulatory reasons: many states only permit financial guarantees to be written by insurers that do no other type of business[343].

The first wrapped municipal bond was issued in 1971: the municipality of Greater Juneau, Alaska, issued a bond to pay for a medical building. This was guaranteed by Ambac, a monoline. Soon a competitor, MBIA, joined the party, and these two companies grew to be the two largest municipal bond wrappers[344].

### 9.1.4  Ambac and MBIA before Structured Finance

The year 2000 is a good place to review the progress of the monolines. At this point, as figure 9.1 indicates, most of the risk of the monolines was municipal bonds in the United States. They had begun to diversify both internationally and into the guarantee of structured finance bonds, but this was not yet a substantial part of their activities.

| | Ambac | MBIA |
|---|---|---|
| **Notional of** **Net guarantees in force** | | |
| Municipal Finance | $180B | $499B |
| Structured finance | $64.6B | $134B (U.S.), $37.5B (Int.) |
| International finance | $31.3B | $10.4B (Int. infrastructure) |
| **Strength** | | |
| Claims paying resources | $5.9B | $9.1B |
| Net Income | $366M | $529M |
| Stockholders' equity | $2.6B | $4.2B |

**Figure 9.1:** Some data for Ambac and MBIA in 2000

At first sight the leverage of both companies seems to be large. MBIA, for instance, had claims paying resources of nine billion dollars supporting a portfolio over seventy five times the size. However its loss experience was small – only three basis points – and the municipal business was both stable and profitable. So perhaps things are not too bad at this point. Cities and states are more likely to default in a recession, it is true, but such defaults were still very rare.

Observe that a monoline writing municipal bond wraps is essentially monetising:

- Their ability to assess muni risk; and

- Their high credit rating.

This business is profitable, not least because the barriers to entry were quite high. It was impossible to be a monoline without a high rating, and a high rating requires a lot of capital.

## 9.2  The Monolines Do Structured Finance

The monolines would probably be thriving AAA-rated companies today if they had stuck to writing financial guarantees on municipal bonds. Sadly for their shareholders, however, they diversified into other business. During the 1990s they began writing financial guarantees on asset backed securities, and this business grew significantly during the Boom years. In particular the monolines began to take on more ABS risk by wrapping the tranches of securitisations.

### 9.2.1  New Business

The attractions of wrapping ABS were obvious. The monolines could diversify away from their traditional concentration on municipal risk into other areas including mortgage risk, corporate credit risk, auto loan risk and catastrophe risk. They did not have to raise money to take on this extra risk: rather in a financial guarantee they are paid at the start of the transaction, and they only have to find cash if the issuer does not pay. Thus the more bonds they wrapped, the more cash they got. Even better, their high credit ratings meant that the monolines could charge substantial fees for their financial guarantees.

The monolines did not, however, charge as much as a credit derivatives counterparty would. In other words, financial guarantees often looked cheap compared with other forms of credit enhancement. There were three main reasons for this, all familiar from the case of AIG discussed in the previous chapter:

1.  Financial guarantees are insurance, so the monolines did not suffer any earnings volatility on the contracts that they had written until losses actually started to happen. Thanks to their accounting, the monolines did not need to support mark to market volatility with capital, and so they did not need to earn a return on the capital deployed for this purpose. This can be seen as an *accounting arbitrage*.

2.  The monolines often did not have to pledge any collateral against the contract that they had written, or only had to pledge it were they to be downgraded. Thus while the monolines remained AAA-rated, they did not need much access to collateral.

3.  Like AIG, the monolines genuinely believed that the losses on the ABS they guaranteed would be small. Their models were calibrated to the past, and these models indicated that they would have to pay out much less than they took in premiums.

The monolines wrote large amounts of financial guarantees on ABS. These were both in the form of wraps of publicly-traded bonds (where the monoline guaranteed the bond whoever held it), and bilateral agreements to guarantee a particular bond for a particular counterparty. The results of updating figure 9.1 to 2007 are shown in figure 9.2. Both companies' American public finance portfolios have increased a little, but the major growth is in structured finance and in non-U.S. risk.

|  | Ambac | MBIA |
|---|---|---|
| **Net guarantees in force** | | |
| Municipal Finance | $281B | $674B |
| Structured finance | $170.7B | $194B (U.S.), $89.6B (Int.) |
| International finance | $72.4B | $64.8B (Int. public finance) |
| **Strength** | | |
| Claims paying resources | $14.5B | $14.5B |
| Net Income | -$3.2B | -$1.9B |
| Stockholders' funds | $2.3B | $3.6B |

**Figure 9.2**: Some data for Ambac and MBIA in 2007

Two trends emerge from the figures, neither good. First, stockholders' funds at both firms did not increase nearly as fast as the notionals at risk. For Ambac, stockholders' equity even fell. Thus both firms became more leveraged. The second trend is the shift from profit to loss. To lose a couple of billion in a year is unfortunate for any firm: to lose a couple of billion when your capital is only a couple of billion suggests that bankruptcy is entirely possible. As MBIA's chairman and CEO wrote in the 2007 Annual Report to his shareholders '2007 ... was not a good year for your company'.

### 9.2.2   Other Exposure

The growth in ABS wraps would be bad enough if that were the only exposures that the monolines had to structured finance. It wasn't. They also took risk by writing credit derivatives on ABS and on ordinary corporate bonds, like AIG. MBIA, for instance, had a notional of more than $200B of credit derivatives in 2007. Moreover some of the premiums received by the monolines were invested in ABS too, so they had exposure on both sides of the balance sheet.

### 9.2.3   Investment Banks and Monolines

The monolines acquired such significant portfolios of ABS risk in part because they ran the best game in town. The premiums that they were willing to write financial guarantees for were often noticeably less than the credit spread of the unwrapped risk, so investment banks could wrap a bond and make a profit. For instance a single A-rated tranche of a mortgage-backed securitisation might trade a credit spread of 40 b.p.s, but a wrap on it would only cost 20 b.p.s., and the wrapped paper – now AAA-rated – would trade at 10 b.p.s. Therefore a bank could make a spread of 10 b.p.s. simply by having the bond wrapped.

One might argue that the spread between the cost of a financial guarantee and the credit spread of the bond is justified, to some extent at least, by the preferential accounting treatment of the wrap; the lower liquidity risk for the wrap provider; and the contingent credit risk the bond buyer was taking. Still, the gap was large enough to puzzle some observers. The monolines seemed to be taking risk for less than the market thought it was worth.

This phenomenon was even more obvious in negative basis trades. Insurance companies were the main providers of CDS protection on ABS, and the monolines were prominent here[345]. Often it seemed that this protection was priced rather similarly to a financial guarantee: the monolines did not seem to charge much for the fact that the CDS was mark-to-market (and that sometimes collateral was required)[346].

### 9.2.4   Counterparty Risk

The monolines' whole business model relied on their high credit rating. Without it, few municipal issuers would be prepared to pay for their bond wraps, and they would be much less attractive writers of protection on structured finance. This is a *cliff risk*: with a AAA rating, the monolines have a business. They can continue writing new wraps and thus making a profit. Without it, they don't and can't. That means that if the monolines suffer losses that are big enough to cause a downgrade then they may well be unable to recover. The ratings agencies are also stranded on the same cliff. Even a one notch downgrade from AAA causes huge problems for a large monoline. If the agency downgrades a monoline without being really sure that it is no longer AAA, then they have effectively pushed a perfectly good firm over the edge. But if they do not downgrade a failing monoline, they lose their credibility with the market.

What might a downgrade mean to a monoline's counterparties? There are several effects:

- Bonds the monoline has wrapped are no longer rated AAA. Their prices fall. Many investors do not have the skills to analyse the underlying credit of these bonds – which is why the wrap was purchased in the first place – so these bonds become illiquid and may well trade below the fair value for a comparable but well understood credit.

- The counterparties of the monolines now have to acknowledge that there is a chance that the protection they benefit from will not be effective. This causes them to write down the value of the protection. Mortgages are often thirty year instruments, so wrapped MBS are thirty year securities. Credit risk on even a AA-rated credit for thirty years is significant, so these counterparty risk related write downs are large[347].

- Some counterparties try to hedge their exposure to the monoline in the CDS market. This may well cause increases in the CDS spread, alerting other market participants to the declining credit quality of the insurer. If there are not enough sellers of protection, this can cause a vicious circle of CDS spread widening, further hedging, and further spread widening.

Thus a big monoline losing its AAA-rating is a very serious affair. It causes losses for the buyers of bonds they have wrapped and for their counterparties. It also hampers the monoline's ability to do new business. It makes raising capital and getting funding more difficult.

### 9.2.5   Credit Support Default

Suppose an investment bank engages in a negative basis trade with a monoline. Conscious of the risk of a downgrade, the bank persuades the monoline to agree to a collateral on downgrade agreement. That is, the monoline agrees that if it is downgraded, then it will post collateral on the CDS it has written. The investment bank therefore considers its counterparty risk to be minimal.

It isn't, though. Let's follow the logic. The monoline is downgraded and so has to post collateral. It does not have sufficient liquidity, so it does not put up the money. The bank can now declare a credit support default (as discussed in section 8.1.6). But should it? If the bank forces default, then it has a claim against the monoline equal to the mark to market of the default swap at the point of bankruptcy. It also has the security the CDS is protecting, and this must now be sold. Providing the security can be sold at once and the bank can extract an amount close to the full value of its claim from the monoline, it can escape unscathed. But if the security is illiquid and its value on sale could be lower than where it is marked, or if the bank is likely to only get a small fraction of the value of the CDS from a monoline bankruptcy, then the case is less clear. Perhaps it might be better to let the monoline carry on doing business rather than push it into default.

This was the dilemma that some firms faced with ACA. ACA was a small monoline insurer which some financial institutions had used as the protection provider in negative basis trades. As the underlying bonds fell in value, the CDS ACA had written became more valuable to ACA's counterparties, and so they called for collateral. ACA had significant losses and was downgraded. It could not post the collateral demanded because it could not raise money[348]. The banks could have forced a default, but they chose not to[349].

ACA approached its counterparties and proposed a forbearance agreement. The counterparties agreed to give ACA some time to negotiate a settlement with them, presumably taking the view that their losses would be smaller in this event than if ACA had been immediately wound up. An agreement was eventually concluded with ACA agreeing to write no new business, make some cash payments, and distribute funds from its existing portfolio to its counterparties[350]. Meanwhile ACA's counterparties had large losses on the protection that they had bought.

In retrospect it is a little unclear why anyone thought that protection written by ACA was effective. The company was not even AAA-rated when the protection was written: its highest ever rating was single A. ACA wrote $69B of CDS protection, $22B of it on subprime, with only $430M of shareholder's funds[351]. CDOs and other types of exotic securities amounted to 90% of its portfolio[352]. Clearly if ACA's counterparties had substantial claims on the policies they had bought, then ACA would have a great deal of difficulty in meeting them. In other words, ACA was only good credit quality provided that no one needed to make claims. Perhaps the firm's counterparties knew that protection bought from ACA had little value and it was purchased as a cosmetic instrument to make their books appear hedged. Or perhaps they failed in their counterparty due diligence. In any event, the case of ACA demonstrates that having a negative basis trade with a collateral agreement is not necessarily a fully hedged position.

### 9.2.6    Quick, quick, slow

Broker/dealer failures happen quickly, as we saw with Lehman and Bear Stearns. Banks also tend to fail fast, at least if they are reliant on the wholesale funding markets. But a pure financial guarantee insurer fails slowly. This is because demands for cash on it are slow and extended: it only has to make good payments on the bonds it has wrapped as they become due. A mispriced portfolio of wraps leads to a slow motion crash rather than a sudden fall.

So what of MBIA and Ambac, the two largest monolines? As we saw earlier, they wrote significant amounts of protection on ABS, mostly in the form of financial guarantees. By the end of 2007, it was clear that the underlying assets of the ABS were not performing, so large claims were likely on the guarantees that the two firms had written.

Concerns at the losses, leverage, and risk of the big two monolines caused the ratings agency Fitch to move first. It downgraded Ambac from AAA to AA on the 19th January 2008, and MBIA on 4th April. Moody's and S&P followed in June. Over a trillion dollars of bonds that the two had wrapped were downgraded at the same time. This was the largest downgrade from AAA in bond market history[353].

These downgrades had a substantial impact. Bonds wrapped by the monolines – including many municipal bonds – fell in value. Liquidity in wrapped bonds dropped. And many municipalities found it harder and more expensive to raise money. The loss of their AAAs did not just hurt the monolines themselves: it hurt a wide range of investors and issuers.

### 9.2.7    Case Study: Bill Ackman vs. the Monolines

William Ackman is a hedge fund manager. He is the founder of Pershing Square Capital Management, a fund which had over $2B under management in 2007. Ackman's style is activist: he tries to unlock value in the firms he invests in.

In 2002, Ackman thought that he saw an opportunity with the monolines. His research[354] indicated that MBIA's business model depended critically on it retaining its AAA rating (for the reasons we discussed above). But he doubted whether the firm really merited its rating, citing (amongst other things) high leverage, the ability to absorb less than $2B of losses before a downgrade, low levels of reserves, and the firm's substantial exposure to structured finance risk.

Ackman displayed admirable patience. He remained sceptical of the monolines business model through the Boom years. Presumably he took positions on that basis, either selling stock short, buying CDS protection, or

otherwise betting on a decline in one or more monolines. Nothing happened
for some years, yet Ackman persisted in his criticism.

Two documents published by Pershing Square in January 2008 give a
further insight into Ackman's thinking.

The first is a letter to the ratings agencies[355]. In it, Ackman questions
the credit quality of the monolines. He claims that they had by then violated
covenants causing them to lose access to liquidity; that their investment portfolios were riskier than they appeared; that their exposure to both residential
and commercial mortgages was significant; and that a new entrant to the municipal finance market – Berkshire Hathaway – posed a significant threat to
their business. One of the final paragraphs of the letter reads:

> Does a company deserve your highest Triple A rating whose
> stock price has declined 90%, has cut its dividend, is scram-
> bling to raise capital, completed a partial financing at 14%
> interest (now trading at a 20% yield one week later), has
> incurred losses massively in excess of its promised zero-loss
> expectations wiping out more than half of book value, with
> Berkshire Hathaway as a new competitor, having lost access
> to its only liquidity facility, and having concealed material
> information from the marketplace?

It may be that some parts of Ackman's critique go too far. But the basic
thrust of his analysis was correct. There were reasons to be seriously concerned about the credit quality of MBIA and Ambac by late 2007.

The second document shows the depth of Ackman's research. It is a letter
to insurance regulators and the SEC, detailing Pershing's research into the
monolines' portfolios of ABS risk[356]. In it Ackman discloses the result of his
analysis of the two firms' portfolios. This analysis suggests that the two firms
are likely to suffer losses far in excess of their capital base. But he also gave
the model away freely on the web so that others could repeat his analysis and
examine the impact of varying assumptions on the modelled losses.

The veracity of the analysis done by Pershing Square eventually earned
Bill Ackman and his investors a lot of money. The share price of the two big
monolines fell and their credit spreads increased through 2007: the downgrades in 2008 sealed the fate of MBIA and Ambac. More than that, though,
Ackman had demonstrated that a smart investor could do a better job at
analysing a portfolio of risk than the ratings agencies or the risk control staff
of some large insurance companies.

### 9.2.8   Get Shorting

The example of Ackman shows how important it is for investors to be allowed to take negative as well as positive views on a company. Without the ability to bet on a share price decrease or a credit spread increase, Pershing Square could not have made money from their analysis.

There are two main ways in which investors can take a pessimistic position. The first is to sell shares *short*, that is to sell shares that they do not own in the hope of buying them back more cheaply later. Typically these shares are borrowed from an investor who does own them, and who is willing to lend them out for a fee[357].

The second approach is to buy CDS protection on a company without owning its debt. This position will be worth more if the company's credit spread increases, therefore the holder will profit if the company's credit quality falls.

Both of these types of position taking make the markets more efficient. Without investors like Ackman, the difficulties at the monolines might have taken longer to uncover, and in the meantime more bond issuers would have purchased wraps. Betting on failure might seem negative, but it has a rôle to play in keeping companies honest.

## 9.3   Insurers and Finance: A Toxic Mix?

Actuaries studying financial risk are sometimes known as type 3 actuaries[358]. Financial risk is in some ways compellingly similar to insurance risk: there is a lot of history, so actuarial modelling is possible, and the mathematics is somewhat similar. So it is no surprise that insurance companies thought that they had the skills and understanding necessary to model financial risk.

It is not true that all the insurance companies who became involved in financial risk lost a lot of money. But many of them did. There are too many examples for bad luck to be a plausible explanation[359]. So what went so wrong for some of the type 3 actuaries and the firms they worked for?

### 9.3.1   The Markets are Visible and Everyone's Actions Matter

Insurance companies wait. That is, once you sell a conventional property, casualty or life insurance policy, you wait to see if there is a claim on it. It is only after the policy has expired that the company finds out what total value of claims have been presented[360]. There is typically little visibility of the risk that has been have taken while the insurance is in force.

The capital markets are not like that. The price of ABS are often visible. Therefore if you have written a financial guarantee on ABS, there is new information every day on the quality of that risk. Moreover your actions in writing protection on that ABS may alter the future price dynamics, helping to inflate a bubble: the behaviour of a hurricane is not altered by the presence of insurance[361].

Insurance companies do not have a culture of mark to market. They may have underestimated the importance of the revised risk estimates provided by market prices as information in their own analysis and, crucially, as drivers of collateral calls or other assessments of their credit quality. Used to operating in the dark, they may have found the light of the capital markets blinding. They may also have underestimated the importance of their own actions in driving the market. Certainly there is little evidence of insurers hedging their exposure to subprime in 2006 or early 2007 as some banks and broker/dealers did.

### 9.3.2   Diversification is Hard to Get

Diversification can be effective in reducing risk in traditional insurance. Thousands of catastrophe policies written in different countries on different risks are diversified since it is hard to think of something that could cause many of these policies to suffer claims at once. Thousands of wraps on many different ABS are not diversified in the same way since investors can withdraw capital from a whole class of bonds simultaneously. If the global economy sinks into a deep, long recession, then all ABS fall in price, regardless of the diversification of the underlying collateral.

The presence of this kind of systematic risk factor which can affect many different types of financial risk simultaneously makes diversification much harder to get in financial risk than in traditional insurance risk. It is also easy to think that you have it, since diversification is present provided there isn't a crisis.

AIG FP provides a good example of the mis-estimation of diversification benefits. The firm used a model to estimate the probability of *any* payout on its CDS portfolio. One account suggests that they estimated there was a 99.85% probability of no pay-out[362]. Yet probability estimates like these depend critically on the correlation between risks in the portfolio. Even a small chance of lots of bad things happening at once can make a dramatic difference to the model's estimates. AIG's model evidently assumed that there was nothing that could cause many of the swaps it had written to become less valuable to it (i.e. more valuable to the protection buyer) simultaneously. That is, it assumed a Credit Crunch could not happen. Only under this assumption was its portfolio well diversified.

### 9.3.3 The Details Matter

There is a temptation amongst mathematically able analysts not to delve too deeply into the details of assets. There are models to build and distributions to fit, after all. But if you are willing to wade through hundreds of thousands of pages, then there is much to be learned from a careful analysis of the asset pool that backs a security. For subprime ABS, one must start with the underlying mortgages with all of their PTIs, FICOs, NAVs, zip codes and so on. In a CDO of ABS, there are the other types of assets, too, all with their own data: credit cards, auto loans, student loans, and so on. Once you have understood the likelihood of receiving cashflows from each of these individual assets, you can build a model of the cashflows of the ABS backed by them and, from there, of the cashflows of a tranche of a CDO built on those ABS. As Ackman's public source analysis of the monolines' portfolios indicates, this is an enormous amount of work. Perhaps some type 3 actuaries did not go this far, starting instead at a higher level of detail. Certainly it seems that those investors that did delve deeply into the assets backing subprime securitisations tended to be short rather than long. Did the insurers' risk analysts review the loan files before they allowed their firms to take mortgage risk?

### 9.3.4 Possible Mistakes

In summary, it seems that there were three types of mistakes that some insurance companies made in their financial risk assessment:

1. Using the past to model the future involves an inductive hypothesis as we discussed in section 6.4.3. This may not be justified[363].

2. Sometimes the market prices of assets were ignored in favour of the firm's own beliefs. Whatever you think of a market price, simply ignoring it (rather than attempting to understand what it is and is not telling you about the fundamental long term value of an asset or liability) is foolish.

3. Overarching both of these errors, there was a kind of group think, whereby the insurers convinced themselves that the models and methods that worked well for their traditional insurance business would also work well in finance. In a sense, they had a hammer, so everything looked like a nail.

Perhaps these generalisation are unfair. But it does seem as though the insurers ended up holding risk that the banking system did not want while getting paid rather less than the banking system thought it was worth.

### 9.3.5   *It's Market Confidence, Stupid*

Finally, note that the long term only matters if you can make it through the short term. It may even be that the monolines will survive: that they will have enough capital for the risks that they have taken. All the same, their shareholders have suffered large losses: offsetting gains in the future are unlikely as it is very difficult for the monolines to write new business without being AAA-rated. The monolines needed not just to be right about their risks. It was also important that the market *believed* that they were right about their risks[364]. The monolines under-estimated the likelihood of a self-reinforcing slide in confidence, share price, credit spread, and access to liquidity that even relatively minor losses could create.

## 9.4   Auction Rate Securities

Floating rate mortgages are cheaper to service on average than fixed rate ones because the yield curve usually slopes up. Borrowers prefer to pay less, all other things being equal, so it is hardly surprising that municipalities sometimes want to issue floating rate bonds.

One of the structures that evolved to service this need is the *auction rate security* or ARS. These have a peculiar nature:

- They are long term securities, sometimes thirty year instruments (or even undated);

- The interest rate paid to investors resets periodically. The period between resets varies, but it is often a week or a month.

- The rate for a given period is determined by an auction. In this auction, the holders of the securities submit bids for the rate which they require to continue holding the security. Investors who wish to buy the security submit bids too, as do those wishing to sell. The auction is Dutch style: rates start high and descend until all the debt has been bought, with all the investors receiving the clearing rate.

Thus far, the situation is not too troubling. The use of an auction to set rates – rather than setting them as a fixed spread to an index such as Libor – is perhaps peculiar. What made ARS dangerous?

### 9.4.1   Auction Rate Security Quality and Market

ARS were intended for investors who do not want to take much risk. The regular auctions mean that they can probably sell at par within a few weeks. But what about the credit risk? The answer was the same as for fixed rate municipal bonds: the issuers got a wrap. Therefore ARS were typically AAA-rated wrapped securities.

The ARS market was significant: at the end of 2007, there were about $330B of securities in issuance. The biggest issuers were municipals and other public sector agencies: the student loan agencies in particular were significant issuers[365].

ARS auctions were arranged by the dealers. These firms underwrote the initial issuance of the paper, then conducted the auctions. Typically the large ARS dealers were either banks or broker/dealers.

There is one feature of the ARS auction process that is worth noting. These dealers did not just arrange the auctions: they sometimes participated in them. That is, the dealers put bids into their own auctions in order to ensure that the auction would clear. This was not illegal, but it does suggest a conflict of interest. On the one hand, the dealers wished to earn the intoxicating fees available from issuing ARS and conducting auctions[366]. They needed to be seen to be able to conduct successful auctions to attract new business and earn new fees. But on the other hand, supporting auctions required the dealers to act as a buyer of last resort, ballooning their balance sheets with ARS that no one wants[367].

### 9.4.2   The Possibility of Auction Failures

An ARS auction can fail if there are not enough buyers. In this situation, those wishing to sell are not able to do so. The dealers were not obliged to bid, so this was a distinct possibility. Investors were compensated for this liquidity risk by a *penalty rate*: if the auction failed, they received a high fixed rate. The penalty rate varied from security to security, but it was often in the teens or even as high as 20%.

The ARS market worked well until late 2007. Investors did not have guaranteed liquidity, but in practice auctions did not fail, thanks to the dealers intervening when necessary. Issuers paid rather lower costs than they would have done to issue fixed rate securities. And the dealers earned big fees.

### 9.4.3   The Reality of Auction Failures

The Crunch came to the ARS market in early 2008. Two factors intruded: questions over the credit quality of the monolines made investors nervous about the quality of wrapped ARS; and the combination of losses and funding stress at many of the dealers made them reluctant to commit balance sheet to supporting auctions.

Nervous investors tried to sell as the auctions for their securities came up. There were insufficient buyers and the dealers did not intervene, so many auctions failed[368]. Investors who thought that they had bought a very high credit quality, liquid security with a low yield found themselves holding a lower credit quality illiquid security paying a high yield. Issuers who thought that their debt service costs would be close to short term interest rates instead found themselves paying a penalty rate which was much higher.

For investors the clue should have been in the return. In the good times, ARS yielded more than money market rates, yet they were often AAA-rated. The capital markets are not known for their generosity, so investors should have known this extra yield was compensation for taking risk. In this case it was for the risk that the auction would fail.

### 9.4.4   Lies, Damned Lies, and Marketing

It may be that some investors who bought ARS were misled. There was certainly the suspicion that this was true, and there were various state and federal probes of the ARS market after the disruption of early 2008. These eventually resulted in a settlement between many of the dealers and the authorities whereby the dealers agreed to buy back some of the ARS they had sold, pay fines, and in some cases compensate investors for losses[369]. Figure 9.3 summarises some of the larger agreements.

| Firm | Approximate Amount of ARS repurchased | Fine paid |
|---|---|---|
| Citigroup | $7.5B | $100M |
| JP Morgan | $3B | $25M |
| Morgan Stanley | $4.5B | $35M |
| UBS | $20B | $150M |
| Wachovia | $9B | $50M |

Figure 9.3: Some settlements relating to Auction Rate Securities

The flaw in the ARS market had always been there. It was an assumption of liquidity. If there were enough investors for the auction to succeed, then all would be well. If the dealer intervened when there were insufficient buyers, then things would still be fine. Investor confidence relied on the quality of the wrap provider, and dealer intervention relied on the availability of risk capital. Both suffered problems in early 2008 and so the ARS market collapsed.

## The Non-bank Banking Crunch

Who bought the municipal bonds and the ARS, the ABCP and the AAA-rated tranches of securitisations? Often it was relatively risk-averse investors. When these investors started to sell, it produced a self-reinforcing cycle of falling prices, illiquidity, declining confidence, losses, capital erosion, margin increases and further liquidation. This process reduced the funding available to new assets. Firms that had sold certainly were not about to put their money back into structured finance products, so assets that would have been financed that way now had to compete with everything else for scarce funding. There had been, if you like, a run on structured finance[370].

This structured finance banking system had few of the aids to financial stability that characterise the conventional banking system: no deposit protection; no constraints on leverage; little disclosure; and few restrictions on what could be sold. It was therefore more susceptible to a run than the conventional banking system. That was to a certain extent understandable: the direct losers when this system went haywire were nearly all professionals who should have known what they were doing. There were severe consequences for the broader economy when many of the assets that had been in funded using structured finance techniques suddenly needed bank finance, especially given this was at the same time as banks were losing money and finding their own access to liquidity was challenged. Thus the failure of the structured finance system was responsible for intensifying the Credit Crunch.

## Notes

[341] See Chapter 3 of Robert Niesley's *Model Jury Instructions for Surety Cases* (Miller Act Payment and Performance Bonds by John Hayes) for more details.

[342] This might be a rating agency. Municipal finance typically takes a different approach because of tax. Muni bonds are tax-advantaged for some retail investors, but many municipalities are not of the highest credit quality. There is strong demand for highly rated instruments from retail investors, so municipalities are (or at least were) willing to pay both for the analysis and for subsequent credit enhancement.

[343] The most important regulation is New York's financial guarantee law. This defines the scope of permitted financial guarantee insurance and governs the conduct of business of all financial guarantors licensed to do business in the state of New York. New York financial guarantee insurers are required to have a minimum amount of capital and to maintain certain levels of reserves. For more details on the regulation of insurance in the U.S., see Kenneth Meier's *The Political Economy of Regulation: The Case of Insurance* (1988), and for general background on the monolines, see the Association of Financial Guarantee Insurers website at www.afgi.org.

[344] For more details of the two companies, see respectively www.ambac.com and www.mbia.com. Two other monolines, FGIC and FSA, also have significant portfolios of risk: there are also a number of smaller players.

[345] Paul Davies, writing in a Financial Times article *New danger appears on the monoline horizon* of 6th February 2008, discusses the issue. He quotes Bob McKee, an analyst at Independent Strategy, as saying that up to $150B worth of CDO business done by the monolines could be negative basis trades.

[346] Part of this may have been because much of the CDS protection written by the monolines was *transformed*. That is, the monoline wrote a financial guarantee to an SPV, and the SPV wrote a default swap to the investment banks. Thus the monoline acted as if it had written an insurance contract, and accounted for the risk as insurance; while the investment bank acted as if it had a derivative, and used fair value. It is also worth noting that the credit derivative was often in *pay as you go* form so it had the liquidity and counterparty risk characteristics of a financial guarantee. For a further discussion of these practices and other features of the monoline's negative basis trading, see Craig Stein and David Aron's *Negative Basis Trade Basics*, Derivatives Week (2006) available at www.srz.com/files/News/f1f91365-9c5d-4052-ac99-07dc5bb58ed9/Presentation/NewsAttachment/a9279aae-c43b-4ee5-b847-7c3a3482ddc2/filesfilesdw.LCCraig.reprint.pdf.

[347] In a Barclays Capital research note of 25th January 2008 *Monolines: The current state of play*, Seth Glasser et al. estimated that monoline downgrades would cause write-downs for banks of at least $20B.

[348] For more details, see Allison Pyburn's article in the Financial Times *ACA's downgrade threat could leave CDS counterparties without recourse* (9th November 2007) available at www.ft.com/cms/s/2/57f389b0-8f0d-11dc-87ee-0000779fd2ac.html. As ACA's 10-Q of 30th September 2007 put it, 'should S&P ultimately downgrade ACA Financial Guaranty's financial strength rating below A-, under the existing terms of the Company's insured credit swap transactions, the Company would be required to post collateral based on the fair value of the insured credit swaps as of the date of posting. The failure to post collateral would be an event of default, resulting in a termination payment in an amount approximately equal to the collateral call. This termination payment would give rise to a claim under the related ACA Financial Guaranty insurance policy. Based on current fair values, neither the Company nor ACA Financial Guaranty would have the ability to post such collateral or make such termination payments.'

[349] The matter is slightly more delicate that that because the CDS were often written by a subsidiary of the holding company. The monoline's counterparties could have forced a default of the subsidiary which had written the protection. This in turn would have caused issues for the holding company. However the operating insurance company (typically a different subsidiary of the holding company, and one with most of the money) often has some measure

of regulatory protection. Thus forcing a default of the CDS writer does not necessarily trigger bankruptcy of the insurance company.

[350] For a full discussion, see the Maryland Insurance Administration Order on the Restructure of ACA available at www.aca.com/faqs/MIAOrder.PDF.

[351] This figure comes from ACA Capital's 10-Q of 30th September 2006.

[352] See David Henry and Matthew Goldstein's story *Death of a Bond Insurer*, Business Week, 2008.

[353] The figure comes from a Bloomberg story by Christine Richard of 5th June 2008, *MBIA, Ambac, $1 Trillion of Debt, Lose S&P AAA Rating* available at www.bloomberg.com/apps/news?pid=newsarchive&sid=a68tNUNifwBg.

[354] See Gotham Partners *Is MBIA Triple A?* (2002).

[355] See William Ackman, *Re: Bond Insurer Ratings*, letter of 18th January 2008.

[356] See William Ackman, *Re: Bond Insurer Transparency; Open Source Research*, letter of 30th January 2008.

[357] Selling shares that you do not own is known as *naked shorting*. This is a disreputable practice. It is even illegal in some countries because it effectively creates new shares. To see this, consider a company with a million shares in issuance, each trading at one dollar. The company is worth a million dollars. A naked short of a quarter of a million shares effectively means that there are 1.25M shares on the market: the real ones and the naked short. The company's worth is now divided among this larger number of shares and hence the share price falls to eighty cents without any change in fundamentals.

[358] The terminology 'type 3' actuary for someone who specialises in underwriting financial risks in an insurance context is due to Hans Bühlmann. See *On Becoming An Actuary of the Third Kind* by Stephen D'Arcy (Proceedings of the Casualty Actuarial Society 1989), available at www.casact.org/pubs/proceed/proceed89/89045.pdf.

[359] In addition to AIG and the monolines, there was the mis-pricing of equity risk in variable annuity policies – an issue that affected many life insurers – and various pension-related miscalculations.

[360] Although if you have written billions of dollars of Florida windstorm cover and you see a hurricane like Katrina developing in the Atlantic, you might suspect that trouble may be ahead.

[361] The value of property in an area prone to hurricanes may however depend on the availability of insurance: high value property is usually only built where it can be insured.

[362] See Brady Dennis and Robert O'Harrow Jr.'s article in the Washington Post *A Crack in The System* of 30th December 2008.

[363] Thus modelling pensions assuming that equities will grow at 9% a year forever is an error: in this case one that many pensioners are paying for.

[364] The lack of visibility of losses, thanks to insurance accounting, did not help here.

[365] A majority of this paper was also tax-advantaged for the retail investor.

[366] A Bloomberg story of 12th December 2008 by Michael McDonald quotes an email from a UBS executive saying that 'Fees from selling auction-rate securities "intoxicated" Wall Street bankers'. See www.bloomberg.com/apps/news?pid=20601087&sid=aFpi_Xhy5BoA.

[367] For more details of the auction process, see *Description of Merrill Lynch's Auction Rate Securities Practices and Procedures* available via www.ml.com.

[368] The New York Times, in a story by Jenny Anderson and Vikas Bajaj of 15th February 2008, *New Trouble in Auction-Rate Securities* available at www.nytimes.com/2008/02/15/

business/15place.html?ref=business quotes 'nearly 1,000' auctions as having failed in that week alone.

[369] There are a variety of cases, some settled, some ongoing. For a representative example, see SEC Press Release 2008-290 *SEC Finalizes ARS Settlements With Citigroup And UBS, Providing Nearly $30 Billion in Liquidity to Investors*, 11th December 2008 available at www.sec.gov/news/press/2008/2008-290.htm.

[370] Some commentators call the structured finance market 'the shadow banking system'. I dislike the term as it seems to me to have connotations of a bad science fiction movie – behold the evil majesty of *The Shadow* – but it does convey the idea that structured finance allowed a system of credit not intermediated by banks to grow. For a discussion of the size of this non-bank banking system, see Timothy Geithner's speech *Reducing systemic risk in a dynamic financial system* (June 2008) available at www.bis.org/review/r080612b.pdf: he suggests that, at $2.2T, assets in asset-backed commercial paper conduits, structured investment vehicles, ARS and related structures were more than 20% of those in entire banking system. The shadow was a large and fearsome creature.

# Chapter 10

---

## The Rules of the Game: Accounting and Regulation

*The arbitrary rule of a just and enlightened prince is always bad.*
Denis Diderot

---

## Introduction

There are two key sets of rules that financial services companies must obey. *Regulations*, imposed by supervisors, must be complied with. These include rules which determine how much capital firms must have. (This capital requirement is usually determined by the risks that they are running.) Financial institutions must also comply with *accounting standards* in their financial reporting.

We begin with accounting. This may seem a dull topic but, like the procession of a Formula One race suddenly enlivened by rain, it can have flashes of excitement. In particular there is an ongoing conflict between those who favour accrual accounting and those who favour fair value which is every bit as hard fought as a decisive Grand Prix[371]. The debate matters because it determines how financial institutions record their earnings, and hence what information investors get about them.

Another key set of rules concern the accounting of securitisations. The ODM only works if securitisation really gets assets off the originator's balance sheet. The ability to do this is therefore key to the growth of new-style banking. Hence it is worth examining how this process works, and how the accounting rules shaped the development of structured finance.

Next we turn to regulatory capital requirements. These have evolved significantly over the last twenty years, in some ways for the worse. The development of bank capital rules is reviewed, and some of the consequences of the rules in force during the Boom years are discussed[372].

One of the purposes of regulatory capital is to constrain firms' leverage. If this cap works, then firms will tend to keep their leverage at or close to the cap in order to maximise earnings. So it is natural to ask what happens to firms using constant leverage as the business cycle waxes and wanes. The answer depends on their accounting, as we will show, and this gives considerable insight into the risk – and the stock price – of firms predominantly using fair value versus those using accrual.

Investors' distrust of accounting played a major rôle in an earlier crisis: the Lost Decade in Japan. This dismal period is discussed, and compared with the current Credit Crunch and the Great Depression of the 1930s.

## 10.1   Accounting and Why It Matters

Accounting is the measurement and disclosure of financial information about a company. This disclosure – via the publication of freely available *financial statements* – provides one of the main tools for the assessment of companies. Investors, counterparties, analysts and regulators all rely on financial statements to inform them about the health of a company. Amongst other things, these statements disclose how much money has been made and what the company's assets are worth. Often they form the most detailed public disclosure that an institution makes.

It is easy to imagine that preparing accounts is purely mechanical: that no two competent accountants could possibly differ on the value of a company. But the reality is very different. First, the valuation of financial assets is often uncertain[373]. This means that if two perfectly justifiable valuation methodologies are applied to the same set of assets, the resulting fair values can easily differ by hundreds of millions of dollars. Second, even if you did know exactly what everything was worth, how that worth is recorded – and in particular how earnings are recognised – is not always obvious. Sometimes choices are available. The process of preparing accounts is therefore complex, error-prone, and oft-times judgemental. In some ways, then, saying a company has made $10M without saying anything about its accounting practices is like saying that something is 5 long without saying if that is 5 feet, 5 meters or 5 light years.

### 10.1.1   *Accounting Standards and Financial Statements*

Given the diversity of ways that a firm could prepare its accounts, how is it supposed to choose between them? The over-riding principle is that accounts should present a *true and fair* view of the company[374]. This general requirement is then elaborated in a series of accounting standards which govern the preparation of financial statements. They define the rules of the game, setting out the broad principles (and sometimes detailed rules) of accounting.

Companies have to publish their statements regularly, typically annually with quarterly updates. These statements contain at least:

- Details of the company's earnings in the quarter, in its *income statement*;

- A summary of what it owns, how it is funded, and its capital in its *balance sheet*;

- An account of what happened to the cash that passed through its hands, in its *cashflow statement*;

- Additional information in the *footnotes* to the accounts.

There are two major sets of standards: American ones, produced by the Financial Accounting Standards Board (or FASB), and International Financial Reporting Standards produced by the International Accounting Standards Board (or IASB). There are some differences between these two sets of requirements. However, they are (slowly) converging, and what unites them is more important for our purposes than what separates them. We will outline the key shared standards which were important in enabling the conditions which led to the Crunch[375].

### 10.1.2  Fair Value and Accrual

There are two main ways of valuing financial assets. These are:

- *Fair value accounting.* This is based on the estimation of a price at which two knowledgeable, willing parties would engage in a transaction in the current market. Thus a $1M bond might be held at fair value, meaning that as its market price changed, we would recognise profits or losses. Its value on our balance sheet would be its current market value, i.e. higher than $1M if the bond's credit spread had decreased, and less than $1M if it had increased. (We discussed fair value earlier in section 3.2.2.)

- *Accrual accounting.* Here net interest income is recorded as it is earned, with any market information on the price of the asset being ignored. Thus if we have a loan of $1M which is paying a credit spread of 1% a year over our cost of funds, that $10,000 is recognised each year as income, and the loan is an asset valued at $1M. Typically a *loan loss reserve* would be taken for a portfolio of loans based on an estimate of the amount we expect to lose from defaults, and certainly if the loan was non-performing, then we would definitely expect there to be a specific loan loss reserve[376]. Otherwise, though, the loan remains valued at $1M under accrual. (Accrual accounting was discussed earlier in section 4.1.6.)

### 10.1.3   Fair Value vs. Accrual

These two approaches are very different. Think of a mortgage we have made. In the first case, to hold it in our financial statements at fair value would be to assume that market-based estimates provide the best guide to its worth. Specifically we would be valuing it where we think we could sell it. In the second case, we simply assume that, unless the mortgage is not performing, it will keep paying. The current market credit price of the mortgage does not matter. Only what we paid to originate the mortgage and what it is paying us matters.

Fair values are relevant if we ever have to sell the mortgage. But if we are well enough capitalised and well enough funded to hold the mortgage to term, then perhaps accrual reflects the reality of our strategy. After all, we will figure out eventually if the mortgage is good or not.

There are arguments against each approach. The opponents of fair value say that it introduces unnecessary volatility into financial statements. Firms that have no intention of ever selling an asset are forced to recognise losses as the non-default components of the credit spread increase. They suffer earnings volatility unrelated to a change in fundamentals. On the other hand, opponents of accrual point out that it allows firms to delay recognising losses that the market knows – or at least has strong reason to believe – are there. The recent sale of Wachovia Corporation provides an interesting example here. Jonathan Weil has pointed out[377] that Wachovia's balance sheet of 30th September 2008 showed that it had a net worth of $50B, yet on 2nd October the Wachovia board recommended an offer for the firm which valued it at $15B. Since the board would never recommend that shareholders should sell their firm for $35B too little, Weil concludes that Wachovia's balance sheet is 'a farce'.

Historically, most of the assets in most banks were held on an accrual basis. This was partly because there was no market for them, so establishing fair value was difficult. More recently, however, more assets have been accounted for under fair value. This is especially so for ODM institutions that often keep the tranches of securitisations (that is, securities) rather than whole loans: these positions are more naturally marked to market than the original loan.

Different financial institutions account for the same risks in different ways. In particular, fair value accounters recognise losses on assets like packages of mortgages as the market deteriorates, but accrual accounters only lose money when they decide to increase loan loss reserves. Since the latter are subjectively determined, the scope to delay write-downs is considerable. This means that there was a considerable suspicion in 2008 that some banks may have been holding assets in an accrual framework in order to avoid having to report losses that would result in a fair value context[378].

### 10.1.4 Being Open About Uncertainty: The Levels of 157

The FASB have developed a helpful approach to the uncertainty surrounding the fair values of some assets. One of their standards[379] requires that firms disclose the breakdown of their assets in three levels:

- In *Level 1* are assets where a market price is readily available. Level 1 assets, in other words, can readily be marked to market.

- In *Level 2* are assets where a market price is not readily available, but where the asset can be valued using a model calibrated using market prices alone. The valuation of level 2 assets therefore has some model risk, but this is mitigated by the fact that all the model inputs are market observables.

- In *Level 3* are all other assets, i.e. those valued using a model one or more of whose inputs is not a market price. Thus anything valued using an historic property – such as an ABS valued using average default rates – will be in Level 3. The valuation of these assets has considerable model risk.

The requirement that firms divide their assets into the three categories provides useful information to investors on the valuation uncertainty in firms' assets[380].

### 10.1.5 Valuation and the Crunch

For something as seemingly staid as accounting, the fair value vs. accrual debate has attracted considerable passion. Certain banks, notably a number of continental European institutions, seem to view accrual with its attendant earnings smoothing as their birthright. They reject the attempt to inform investors what their assets might be worth were they to be sold. Others equally fervently embrace fair value. The reality is that while fair value works well for liquid instruments, it is much more problematic for illiquid ones. Here it can become almost (but not quite) as judgemental as accrual. The fact that the liquidity of financial instruments can change also complicates the situation. Neither method is perfect. In a crisis, the absence of a method for accounting for most of a financial institution's balance sheet which is widely accepted to be true and fair hardly helps investor confidence.

On balance, though, fair value gives too much valuable information to abandon. Accounts prepared using fair value must be read with an understanding of what fair value measurements relate (and what they don't). But provided the reader understands that fair value does not necessarily equate to long term holding value, then earnings under fair value give more insight than those reported under accrual. As the SEC says in its study on Mark-To-Market Accounting[381]:

> Fair value information is vital in times of stress, and a suspension of this information would weaken investor confidence and result in further instability in the markets. [Market] participants pointed to what they believe are the root causes of the crisis, namely poor lending decisions and inadequate risk management, combined with shortcomings in the current approach to supervision and regulation, rather than accounting. Suspending the use of fair value accounting, these participants warned, would be akin to "shooting the messenger" and hiding from capital providers the true economic condition of a financial institution.

### 10.1.6   Consolidation

When a company issues a bond, that bond is a liability of the company: they have to pay it back. On the other hand, when a securitisation SPV issues a bond, the originator of the assets does not have to pay that bond back: the assets are the only backing for the ABS (the clue being in the name). So the originator (and the sponsor) usually wishes to ensure that the SPV's securitisation tranches do not count as its liabilities. If an SPV *deconsolidates*, then it does not count as a subsidiary of the company: but if the firm is required to *consolidate* an SPV, then any debt the SPV issues will count as a liability of the firm. Consolidation therefore inflates the originator's balance sheet and increases its visible leverage.

The consolidation rules, then, are key to the securitisation market. Issuers wish to ensure that these rules allow them to deconsolidate securitisation SPVs. The topic is however complex: clearly if a sponsor controls the actions of the SPV, or owns the vast majority of the SPV's debt, then the relationship between them can be sufficiently incestuous that deconsolidation is not merited.

### 10.1.7 The Consequences of the Consolidation Rules

This is one of the areas where U.S. accounting standards have been quite helpful to the industry. The FASB permitted two methods of achieving deconsolidation: either ensure that your SPV meets certain criteria (so that it is a 'qualifying SPV'), or structure it so it does not consolidate as a *variable interest entity*[382]. These rules are quite mechanical and offer a safe harbour that is relatively easy to enter. They are therefore responsible for the growth of securitisation in the U.S. The IASB rules, used in Europe and Japan, are based on general principles: deconsolidation can be problematic here. But in the U.S., the recipe for success was clear.

It is easy to criticise various aspects of accounting rules, and certainly specific features of them – such as the safe harbour for securitisations in the U.S. – facilitated the growth of the ODM. But the underlying problems go back much further. Common stock corporations became popular during the Industrial Revolution. This was the first period in history when many firms needed a lot of new capital. Hence share issuance grew. The holders of these shares needed information to allow them to assess the performance of the companies they owned. Thus accounting standards developed to present a true and fair view of 19th century companies. At that time, most companies made things or provided simple services; most companies were funded by a combination of equity (which took risk and had control) and senior debt; and both asset valuation and earnings recognition were relatively simple affairs.

Today, accounting standards still have to deal with companies like this, but they also have to handle banks with over a trillion dollars worth of assets and extensive books of derivatives; zombie SPVs issuing multiple tranches of debt with sophisticated credit enhancement; and originators who own assets for a short time then sell them but retain some interest in (and possibly some control over) their performance. Securities whose character is between that of equity and that of debt are commonplace; control does not necessary go with equity ownership; nor does taking risk require owning an asset. It may be that a framework that worked well for classical corporations is not capable of easy adaptation to the world of structured finance. The IASB and the FASB have formed a Financial Crisis Advisory Group to 'consider financial reporting issues arising from the global financial crisis'. Until it reports, we will just have to make do with the current framework[383].

## 10.2    Regulation and Regulatory Capital

Financial stability is an evident good. But what does it mean exactly? There are a number of aspects which are common to many definitions:

- *Failures must be orderly.* It is unrealistic to expect that financial institutions will never fail. Perhaps it should be rare for large banks, but there are many financials which are not banks, including small specialists in a number of areas such as commodities or investment management[384]. So failures are inevitable, but when they happen, they should not threaten the wider financial system.

- *The safety of (retail) deposits.* The ordinary man and woman cannot realistically be expected to assess the credit quality of banks. And they tend to get upset if they lose their money. Governments therefore try to ensure that retail deposits are safe, by a combination of capital requirements (which make bank failures less likely) and deposit insurance (which protects deposits should a failure occur).

- *Price stability or the control of inflation* and

- *Orderly markets with sufficient asset liquidity.*

(We have already dealt with the last two of these issues in chapters 3 and 4.)

Many of these objectives are threatened by *systemic risk*: the risk that the failure of one firm causes more widespread instability in the financial system. This contagion can be direct, because other firms are owed money by the failing firm. But in the Crunch it was more often indirect, with a failure causing a general reduction in confidence, and forced liquidation of the assets of a failed institution causing losses at other firms due to mark to market.

The presence of systemic risk makes it clear that the financial system is exactly that: *a system*. The coupling between different firms varies, but if it is generally too tight – if firms are too reliant on each other – then systemic risk is high. There are two ways to ensure that this does not happen: ensuring that individual institutions are unlikely to fail; and making sure that no firm is too reliant on another.

Regulatory capital requirements attempt to do this by requiring that financial institutions have enough capital to support (some of) the risks that they are running. Thus they should not fail too often. Moreover, extra capital is imposed if a firm relies too much on another, thus penalising excessive connections in the financial system. That is the theory, anyway. Let us now turn to the practice.

## 10.2.1  A Small Town in Switzerland

In the 1980s, banking was beginning to be an international business. For the first time banks started to lend significant amounts of money outside their home countries. This meant that countries could compete on capital. Consider a Canadian bank lending to American corporates. If Canada's capital requirements had been lower than those in the U.S. for the same risk, then the Canadian bank could take more risk for the same capital, and thus potentially earn more money. American banks would protest, and U.S. capital standards might be lowered to meet Canadian ones. This chase for the bottom is clearly in no country's long term interest, so it became important to agree on international standards for regulatory capital.

The venue for this agreement was Basel, a town in Switzerland. The Basel Committee on Banking Supervision had been established by the Governors of G10 central banks in 1974 at their bank, the Bank of International Settlements in Basel[385].

The Committee managed to agree on a framework for bank capital requirements known as *Basel 1* in 1988: it is sometimes also known as the *1988 Accord*. This required banks to hold capital for the credit risk in their loans books. Indeed, it is from the 1988 Accord that the 8% of notional capital requirement we used in section 1.2.2 came from.

## 10.2.2  Something Simple: The 1988 Accord

In 1988, few people thought that they could quantify *all* the risks run by banks, least of all regulators. The priority was instead to provide a reasonable amount of capital for the most important risk – the credit risk of loans – and to get an agreement. In this sense the 1988 Accord was agreeably modest: it was moreover successful. There was international agreement, and thus common standards for internationally active banks in the G10 were adopted.

Basel 1 subsequently received a good deal of criticism. Most of this was on the grounds that it is not *risk sensitive*. A loan of $100M to a corporate required $8M of capital whether the corporate was AAA or BB rated. Similarly, residential mortgages required 4% of notional; many interbank loans 1.6%, and government loans were free[386]. Within each category the capital was independent of the quality of the loan.

On the face of it, an absence of risk sensitivity seems like a bad thing in capital rules. However, we need to tease apart several threads before commenting on that. There are two aspects to risk sensitivity:

- *Portfolio Risk Sensitivity*. This occurs when more risky portfolios require more capital.

- *Temporal Risk Sensitivity*. This happens when the capital required for the same portfolio changes in time based on market conditions.

Basel 1 has neither kind of risk sensitivity. The lack of portfolio sensitivity means that banks have an incentive to take worse risks rather than better ones: a loan to a good quality borrower has a lower spread than that to a bad borrower, but it requires the same capital. Therefore bad borrowers will tend to be preferred. This however is not too bad since most very good quality corporate borrowers are active securities issuers anyway: they often get their money from the bond markets rather than from banks.

The lack of temporal risk sensitivity is, however, a really good thing. To see why, suppose that Basel 1's capital requirements for corporate loans ranged from 6% when the economy was doing well and not many corporates were defaulting, to 10% in a recession. This is genuinely risk sensitive. It is also terrible for the economy: as a recession arrives, banks' capital requirements go up. They can support fewer loans with the same amount of capital. Thus they restrict lending just as the economy worsens, making the recession longer and deeper than it need be. Thus capital requirements which display temporal risk sensitivity are said to be *procyclical*: they increase when the markets are falling.

### 10.2.3  Market Risk in '96

By the mid-1990s, credit risk was no longer the dominant risk class in some large banks[387]. It became important to standardise capital requirements for market risk too. The 1988 Accord was amended in 1996 with further rules for market risk.

These rules encapsulated two important decisions.

- The 1988 Accord had the same rules for everybody. The 1996 Market Risk Amendment offered several alternatives: a simple set of rules for less sophisticated firms; and a more complex, more risk sensitive (in both senses) method for calculating capital for those banks that could convince their supervisors that they should be allowed to use it. The more complex approaches allowed firms to use their own risk models to calculate capital, so they are sometimes called *internal models* approaches.

- Basel 1 did not articulate why 8% of notional was the right amount of capital to hold. The 1996 Amendment described *how* capital should be calculated as well as *what* amounts were necessary. The idea was roughly that firms should hold sufficient capital to protect themselves against a one in a twenty five year event[388].

These two decisions had consequences. The first decision meant that the same risk no longer attracted the same capital charge in all banks. The internal models approach required much less capital than the simple rules, so banks had an incentive to use it. Bigger, more sophisticated (and thus more systemically risky) institutions had lower capital requirements for the same risk than smaller firms. This is a perverse outcome.

Firms using the internal models approach are permitted to use their own calculations of how much capital was needed for a one in twenty five year event. It was hard to know if these models work, since to be sure you had enough capital for a once in twenty five year event, you would need to wait a hundred years or so. Then, when you had experienced a handful of such events, you would have a good idea of whether your model worked. Of course, the economy changes so fast that any model which requires as little as twenty years observation to validate can never be shown to work[389]. But as it turned out, the financial system only needed eleven years from 1996 before it became clear that many firms' models did not work[390].

### 10.2.4   The Basel 2 Project

The Basel 1 capital requirements for credit risk survived for more than ten years. Despite their crudeness, these rules enhanced the safety of the banking system. Partly this was because they provided a common international language for describing bank safety: investors could compare a bank's total capital with its capital requirements; the more it had compared to its needs, the better. Banks therefore took care to look good under this measure.

The advent of CDOs allowed banks to preserve appearances while increasing their returns. The basic arbitrage was to securitise a portfolio of loans, sell the senior tranches but retain the junior and mezzanine risk. For a $1B portfolio of corporate loans, the senior tranche might be the top $900M of the transaction. Under Basel 1, then, selling this trade would have reduced the exposure from $1B to $100M and hence the capital required from $80M to $8M. The vast majority of the risk, though, is in the bottom tranches. Selling the senior only removes the risk of a catastrophic level of losses: the risk of typical levels of default is retained. Securitisation therefore allowed banks to conduct *regulatory arbitrage*: reducing capital requirements but not significantly reducing risk. They could – perfectly legally – increase their leverage and thus their potential returns while remaining capitally adequate. Regulators had to do something about this, and so the Basel 2 project was born.

### 10.2.5   Designing a New Capital Accord

Basel 2 is an industry. It has occupied thousands of people for over ten years. The first proposals were in 1999[391], and the 'final' framework did not appear until 2004. This however was not really final, and the comprehensive version of Basel 2 was eventually published in 2006[392]. National regulators worked on implementing the Accord. By 2008, some had rolled out the new standards.

The final Basel 2 capital rules are complicated and extensive. They correct many of the perceived deficiencies of Basel 1, producing capital charges that are much more dependent on risk. Some forms of regulatory arbitrage become impossible[393], and asset classes that the supervisors particularly disapproved of around the year 2000 were hard hit. The new capital accord is, however, a failure. It allowed the largest banks to take more risk and thus decreased financial stability. And it made the Crunch more likely. How did the regulators get the new rules so wrong? To understand this, we need to look at how Basel 2's rules work.

### 10.2.6   Calculating Capital

Figure 6.1 (reproduced here as figure 10.1) provides a good illustration of the calculation of capital. It shows the distribution of losses due to defaults on a particular portfolio of 100 five years loans, each of $10M. There is slightly less than a 50% chance of losses exceeding $20M so the *expected loss* is roughly $20M. This is the amount we should reserve against the expected level of defaults on this portfolio.

There is a slightly less than a 1% chance that the losses in the portfolio will exceed $65M. Therefore if we wish to keep sufficient capital to ensure that only a one in a hundred event endangers the bank, then we would allocate sufficient capital to absorb $65M worth of losses. $20M has already been reserved, so the capital requirement for this portfolio at a *confidence interval* of 99% is $45M.

Basel 2 attempts to calculate the capital required for the credit risk that a bank takes at a 99.9% confidence interval and a one year holding period. That is, it requires banks to keep sufficient capital to survive a one in a thousand year event. The advanced models in Basel 2 are known as *internal ratings based approaches* or IRB approaches. They do the calculation as follows:

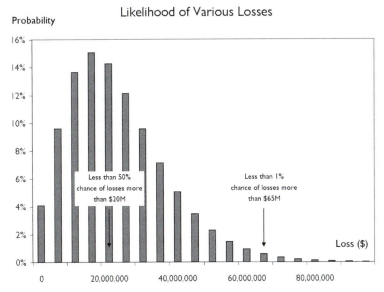

**Figure 10.1**: The probability of various numbers of defaults in a portfolio of 100 loans, with expected loss and 99% unexpected loss levels illustrated

- Banks are permitted to estimate the credit quality of each borrower, their recovery, and the amount that will be owed at default (this last being important because the bank may have offered a line of credit which can be drawn down);

- These parameters, along with the maturity of the loan, go into a model *specified by the supervisors*. This is similar to the model we used to calculate the loss distribution in figure 10.1: its output is the capital required to cover credit risk at 99.9% confidence[394].

In particular, this formula defines the correlation between exposures. Banks are not permitted to estimate this themselves: the supervisors specify it for them[395]. For a large portfolio of credit risk – what most banks have – these hard-wired correlations are a crucial driver of the amount of capital the bank is required to keep for credit risk.

### 10.2.7 Risks included in Basel 2

The New Accord has rules which require capital for credit risk. Most large banks use an IRB approach. They will also typically use an internal models approach for market risk. The final risk that Basel 2 considers worthy of capital is *operational risk*: banks have to keep some additional capital to support earnings volatility coming from this source.

Are these all the risks that matter?

### 10.2.8 *Risks ignored in Basel 2*

Capital requirements alone cannot remove systemic risk. But prudent capital requirements can make the financial system safer, especially if they require a buffer against the most important causes of earning volatility. It is therefore a surprise to discover that the Basel accords do not require capital to put up against four common risks in financial institutions.

1.  There is no extra capital required for *asset illiquidity*: illiquid assets require the same capital as liquid ones;

2.  There is no capital requirement for risk in how the institution is funded. Thus interest rate risk in funding attracts no capital requirements – unlike other types of interest rate risk – and funding liquidity risk is also free[396].

3.  Capital is not required for any business or strategic risks.

4.  Finally, Basel 2 provides a clear route map to minimising capital requirements for securitised assets (including assets in conduits and SIVs). Retained tranches do attract capital. However off balance sheet funding allowed firms to keep much of the return of a portfolio while avoiding the requirement to keep much capital for this additional *structured finance risk*. Off balance sheet funding allowed many institutions to lever themselves up significantly during the Boom years.

The last issue is important, but even without structured finance and the ODM, Basel 2 is not comprehensive. The extent of the gap between the risks covered by the Accord and risks causing losses to traditional banks is assessed in research by Andrew Kuritzkes and Til Schuermann[397]. They decomposed banks' earnings to discover the contributions of various risks. Figure 10.2 summarises their results: it suggests that Basel 2 requires no capital for roughly a third of the risks that conventional banks run. In particular, as a senior British regulator, Paul Sharma, said[398]:

> The comparative absence until recently of liquidity risk from
> the regulatory debate... should surprise us a great deal.

## 10.3 The Consequences of Basel 2

An enormous amount of effort was spent by both regulators and the financial services industry on the new Accord. The vast majority of it was expended

| Risk | Included in Basel 2? | Percentage Contribution |
|---|---|---|
| Market Risk | To some degree[399] | 6% |
| Credit Risk | Yes | 46% |
| Operational Risk | Crudely | 12% |
| Funding Liquidity Risk | No | 18% |
| Business Risk | No | 18% |

**Figure 10.2**: A summary of the coverage of risks by Basel 2 in traditional banking and their contribution to earnings volatility

on the details: should the capital treatment for a particular instrument be this or that? A few commentators noticed that there were bigger problems with the whole conceptual framework of the Accord: these objections were cordially ignored. As a result, Basel 2 had three serious and probably unintended consequences.

### 10.3.1  Bigger Is Better

Basel 2 used the idea of multiple capital techniques for calculating capital originally promulgated in the 1996 Market Risk Amendment. Smaller banks typically used simpler, less risk sensitive capital calculations whereas larger banks used the more advanced methods. These advanced techniques often produced lower capital requirements. This in turn meant that a large bank could buy a small one, and use its advanced technique to calculate capital requirements for the smaller bank's portfolio. The answer would be lower, thus giving an incentive for large banks to acquire smaller ones. Clearly concentrating the risk of the banking system into a small number of large institutions decreases financial stability. As Karen Shaw Petrou said in testimony before the U.S. House of Representatives in 2005[400]:

> If the rules are wrong and unintended competitive consequences ensue – small banks are swallowed by big ones or specialized banks are gobbled by diversified ones – a revision to the rules won't put the banking system back together.

Perhaps too big to fail institutions should be required to hold a higher capital requirement for the same asset than their smaller peers? In any event, Basel 2's encouragement of the growth of large institutions at the expense of smaller ones should not be accepted without question.

### 10.3.2 Procyclicality

The Basel 2 capital rules display both portfolio and temporal risk sensitivity. The more advanced rules in particular are highly risk sensitive, since they rely on banks' own assessments of the credit-worthiness of their counterparties. Thus the rules are procyclical. This effect is not minor: in a recession, capital requirements for a bank's loan portfolio can increase by 30% or 40%[401], reducing the availability of credit just when it is needed most. This intensifies the downturn.

### 10.3.3 And The Winner Is...

The move from Basel 1 to Basel 2 produced some winners and some losers. Firms which were not conventional banks were typically losers: specialists in asset management, corporate finance, or the capital markets typically saw their capital requirements increase when the new rules came into force. Retail banks were the biggest winners as the capital requirements for their activities decreased. In particular, there was one activity which became significantly cheaper under Basel 2: mortgage lending. Thus Basel 2 substantially reduced capital requirements for the business that caused the biggest financial crisis in two generations[402]. A mis-judgement of this magnitude cannot be ignored in any analysis of the success of the New Accord.

### 10.3.4 Northern Rock's Basel 2 Waiver

The Treasury Committee's report into the failure of Northern Rock (discussed in section 4.6) gives a comprehensive discussion of the dangers of the Basel 2 treatment of mortgages[403]. Northern Rock applied for a waiver permitting it to use an internal ratings based approach, and permission was granted by the bank's regulator on 29th June 2007. As the report says, thanks to this waiver

> Northern Rock felt able to announce on 25 July 2007 an increase in its interim dividend of 30.3%. This was because the waiver ... meant that Northern Rock had an "anticipated regulatory capital surplus over the next 3 to 4 years".

In other words, just as the problems in mortgage lending were becoming apparent and strains in the financial system were rising, regulators were *reducing* capital requirements for retail banks. This allowed some banks to take more risk. Specifically, Basel 2 encouraged banks to increase their exposure to retail mortgages at the worst possible time.

### 10.3.5   Model Risk and Internal Models

The direction of bank regulation since 1996 has been towards greater and greater use of models, either banks' own, as in the market risk capital requirements promulgated in the 1996 amendment, or the supervisor-specified credit risk models of Basel 2. This was motivated by desire to better align regulatory capital requirements with banks' own risk management practices, and to make the rules risk sensitive.[404]

Model risk is a big issue in banks' own risk assessment. But it is even more significant when models are used for regulatory purposes. If a bank gets their risk assessment wrong, then they should suffer the consequences. But supervisors are supposed to protect the financial system, so they have a higher duty of prudence. It is therefore particularly troubling that risk models have such a prominent place in regulatory capital calculations. We have already discussed the failures of market risk models earlier (see section 6.6.3): what about the model built into the IRB?

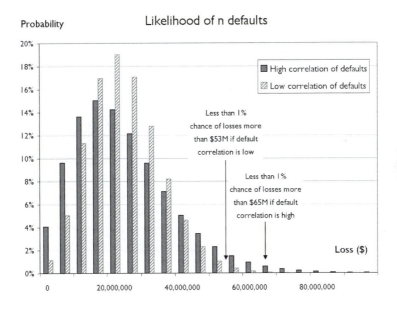

Figure 10.3: The variation of unexpected loss with default correlation

This suffers from model risk too. Specifically, the correlations the supervisors specified are rather generous, especially to retail banking. This is why mortgage banks in particular do well under the New Accord: the Basel Committee calibrated their model to the benign conditions of the Boom years, and hence produced a set of rules that encourages banks to take more mortgage risk[405]. This model risk means that the new Accord in some ways *decreases* financial stability by allowing banks to operate without enough capital to keep the banking system safe. Basel 1 might have been crude, but its prudence does not rely on calibration in the way that the more advanced models do: perhaps it is time to go back to simpler capital rules?[406]

Figure 10.3 illustrates the issue. If default correlations are set too low, capital estimates will be too low too. Moreover the issue gets worse as the confidence interval increases: the mis-estimation at 99% is not too bad, but at Basel 2's 99.9%, the error is much bigger[407].

## 10.4   Regulation Away from Basel

Basel 2 defines the capital requirements for internationally active banks globally. It has also been adopted in the European Union for many other firms[408]. But, crucially, it does not apply to insurance companies (who are regulated under a different set of rules); hedge funds (who typically are not regulated at all); or broker/dealers (who, until they disappeared, were regulated under the SEC's CSE rules[409]). Does this matter?

### 10.4.1   *Diversity and the Level Playing Field*

If capital requirements differ from one class of firms to another, it encourages risk to move. Therefore as insurance companies have lower charges for credit risk than banks, there is an in-built incentive for credit risk to leave the banking system for insurance companies. This is probably not a good thing, especially if you take the view that banks know more about managing credit risk than insurance companies.

The other view is that if you have the same capital rules for everyone and they are not completely correct, then everyone has the same perverse incentive. At least differing regulatory systems give a degree of robustness to the financial system. If you believe that a collection of regulators can write a perfectly correct set of rules which no amount of effort, even from a vast collection of highly paid and talented bankers, can arbitrage, then a single set of rules is good. But if you are not a fantasist, then you would probably consider that a measure of regulatory diversity increases financial stability.

Diversity needs to be managed, though. Regulators must be alert to cross-sectoral risk transfer[410]. They should intervene if it appears that this is producing significant systemic risk.

### 10.4.2    *Monoline Capital Adequacy*

The problem of capital requirements differing between classes of institutions is well illustrated by the monolines[411]. These firms are mostly regulated by the New York State insurance commissioner (insurance is a matter of state rather than Federal regulation in the U.S.). They also have to meet ratings agency requirements for their desired credit rating. In practice these were the more constraining requirements during the Boom years, so the big monolines were effectively constrained by the capital that the ratings agencies required them to have in order to qualify for a AAA rating.

The detail of these rules differs from agency to agency, but the overall design principles are roughly similar: in each case, the model calculates the loss to the insurer for a number of stressful scenarios. In order to be AAA-rated, the insurer must have sufficient capital to remain solvent through an extended period of stress. Unfortunately the ratings agencies' definition of 'stress' did not always involve particularly large market moves. Even as late as 2007, for instance, Standard and Poor's was quoting 0.1% as a representative capital charge for the top ('super AAA') tranche of a CDO[412]. Given that some of these securities had lost 30% or more of their value by the end of 2007, this capital charge did not provide much of a margin of safety.

The problem, then, was that the agencies used exactly the same modelling methodology as the people they were rating. There was little diversity of approach, and little cognizance of model risk. A truly conservative capital model – whether for an insurance company, a bank or a broker/dealer – should assume that *any* non-government bond has a minimum level of risk[413]. It should also severely restrict diversification benefits because, as we have seen, in a crisis all risky assets tend to fall in value together.

## 10.5    Understanding Earnings

Many investors have come to the conclusion that fair value is the worst way of reporting earnings – apart from all the others. Whatever your predilections here, it is important to understand the differences between earnings reported using differing conventions. This is especially true since retained earnings form part of a financial institution's capital, and so they can support further risk taking. Two examples illustrate the situation vividly: we look at the same business accounted for under accrual and under fair value, and show how the same strategy executed using two different earnings measures can produce rather different outcomes.

### 10.5.1  Accrual Accounting, Capital Requirements, and the Behaviour of Financial Institutions

Suppose we have a very simple old-style bank with 100 dollars of capital and 1,200 dollars of long term loans, and consider a year of improving asset quality followed by a year of falling quality. We will look at how the bank's earnings are recorded over time. In particular, consider a good year, followed by a bad year.

Under accrual, earnings are given by net interest income minus loan loss reserves (LLR). These and other key financials for our example bank are shown in figure 10.4: we assume that the bank's profits are split half-and-half into dividends for shareholders and retained earnings, so when the bank makes 20 dollars, it gives 10 to shareholders and retains 10.

|  | Start of Year One | End of Year One | Start of Year Two | End of Year Two | Start of Year Three |
|---|---|---|---|---|---|
| Capital | 100 | 100 | 110 | 110 | 111 |
| Gross Assets | 1200 | 1200 | 1200 | 1320 | 1320 |
| NII before LLR |  | 30 |  | 32 |  |
| LLR |  | (10) | (10) | (40) | (40) |
| Net Earnings |  | 20 |  | 2 |  |
| Dividends paid |  |  | 10 |  | 1 |

**Figure 10.4**: Key financial variables for an old-fashioned bank under accrual accounting

The bank has good net interest income in year one thanks to low expectations of losses. Its capital rises in year two thanks to retained earnings: these now count as part of capital.

Suppose that the bank keeps a constant leverage in order to ensure that shareholders get a good return. This would then cause the bank to originate 120 dollars more of assets in year two. Unfortunately this is a bad year: rising defaults cause an increase of 30 dollars in the LLRs the bank estimates that it needs, and so earnings are much lower that year. (32 of earnings minus a 30 increase in LLRs gives 2 of earnings.) However earnings are still positive, and the bank remains well-capitalised based on a capital requirement of 8% of notional.

Note that under a Basel 2 advanced capital calculation, the capital requirements would be a smaller percentage of notional in year 1, allowing more leverage. In year 2, though, they would be a higher percentage, making it harder for the bank to keep on lending. Procyclical capital requirements are credit crunch generators.

### 10.5.2   All the Fun of the Fair

Now consider the same bank with the same assets but using fair value accounting, perhaps because the assets have been securitised and sold to it rather than originated by the bank itself. The key financials are now as laid out in figure 10.5.

|  | Start of Year One | End of Year One | Start of Year Two | End of Year Two | Start of Year Three |
|---|---|---|---|---|---|
| Capital | 100 | 100 | 125 | 125 | 75 |
| Assets | 1200 | 1200 | 1200 | 1500 | 1450 |
| Earnings |  | 50 |  | (50) |  |
| Dividends paid |  |  | 25 |  | – |

**Figure 10.5:** Key financial variables for a financial institution taking credit risk under fair value accounting

Tightening credit spreads in year one causes greater earnings. This results in greater dividends for shareholders and larger retained earnings. If this bank keeps its leverage constant, then there will be much faster asset growth in year two. *Just because of its accounting* this institution appears to earn more than the old-fashioned bank. These earnings allow it to grow faster. Thus there is larger notional at risk when the bubble bursts in year two. The effect on the firm's earnings is much greater as the full fall in asset fair value hits the P/L rather than just the increase in the default-related component of the spread[414]. The loss leaves the bank badly capitalised: it has roughly 5% of capital to support its risk rather than the 8% required under Basel 1. This means it must either raise new capital or sell assets. Furthermore, the effect would be worse under a more temporally risk sensitive capital regime.

This is of course a stylised example, but it does illustrate the dangers of fair value accounting combined with constant balance sheet leverage[415]. If institutions with a predominance of fair value assets do not decrease leverage in good times, then they are at greater risk than accrual accounters when a recession hits and market values drop[416]. It is therefore vital that the readers of financial statements understand the difference between the lower, smoothed earnings accrual produces and the more volatile, more responsive earnings that come with fair value.

# 10.6   Japan's Lost Decade

Consider the following summary, written after the crisis had passed:

> *The U.S. banking crisis presents a particularly apt opportunity for a case study for three reasons. First, most of its underlying causes (excessive asset expansion in periods of economic boom, liberalization without an appropriate adjustment of the regulatory environment, weak corporate governance and regulatory forbearance when the system is under stress) are typical of banking crises in general. Second, the U.S. banking crisis serves as a warning that such a crisis can befall a seemingly robust and relatively sophisticated financial system. The fact that only a decade ago U.S. banks were considered to be among the strongest in the world makes the extent of their decline all the more remarkable. Finally, the U.S. banking crisis demonstrates that the costs associated with such a crisis can be considerable.*

This account seems accurate: but it was not written in 2009. It in fact comes from a working paper on the Japanese banking crisis of the 1990s by Kanaya and Woo[417]: I have simply substituted 'U.S.' for 'Japanese' in the original text. This legerdemain makes it clear that major financial crises often share features. These parallels will be discussed in the next section. But first, what happened in Japan in the 1990s?

## 10.6.1   *Bubble and Crash: Japan towards the Millennium*

'The Lost Decade' is a term used to refer to the 1990s in Japan[418]. It was an extended period of low or negative economic growth, falling asset prices, and failing companies. The 1980s had seen a period of rising property prices, tightening credit spreads, and rocketing equity markets. The main Japanese equity index, the Nikkei 225, peaked on December 29th, 1989 closing at an all time high of 38,915.87. For the next ten years, the trend was down. Stock prices fell, property prices fell, and Japan slipped into recession. Financial institutions were seriously weakened, and the Bank of Japan, having raised rates in 1990, cut them sharply in an attempt to avert disaster.

The principal cause for such a long period of economic malaise was the extent and treatment of bad loans within Japan's banks. Most banks had lent extensively against property during the late 80s, and many of these loans turned bad as property prices fell. The banks did not admit to these problems readily, and the government was slow to intervene. The first moves came in late 1994. These were, however, both small and unconvincing: rates were cut to close to zero[419] and the Bank of Japan's other interventions were not enough. As the authorities floundered, banks were failing[420]. It was only when all hope of market revival had been extinguished – and thus when confidence in the solvency of Japanese banks was low – that effective steps were taken.

### 10.6.2   Regulation and Accounting in Japan

The reason that many Japanese banks were able to last some years before admitting to their problems was accounting. The problem loans were held in accrual accounted books, and there was no pressure either from management, shareholders or auditors for the banks to record realistic loan loss reserves. Loans were sometimes restructured despite being non-performing: this allowed the banks to avoid admitting that they were unlikely to ever return principal. The desire not to lose face by being the first to admit to serious problems meant that many banks' accounts were essentially fantasies for much of the 1990s. It was only in 1998-99, when realised losses started to build sufficiently to threaten the larger banks, that bank disclosure improved somewhat.

The problem with delaying loss recognition is that it affects confidence. Investors realise that there is a problem. Lacking information, they assume the worst. Thus there is a case to be made that accrual accounting damages financial stability to the extent that investors in financial institutions can question the adequacy of loan loss reserves. Fair value is not a panacea, but it does give fewer places to hide[421].

Another problem in Japan was that banks often held significant equity stakes in their clients. This was a result of the *keiretsu* system: Japan's economy is dominated by large groups of companies, each group including a variety of industrial companies served by one or more banks. Thus for instance the Mitsui *keiretsu* was a group of companies including Fuji film, Mitsui real estate, Mitsukoshi retailing, Suntory whisky, Toyota cars, Toshiba electronics, and both Mitsui and Sakura banks. The financial services needs of the group companies were almost entirely provided by the group banks. The equity stakes banks held in group companies were often held at historical cost and so represented a significant unrealised gain for many banks.

The Japanese implementation of Basel permitted these unrealised gains to be counted as a part of the banks' capital. Thus as the equity markets fell, banks' capital fell too, challenging their capital adequacy. Some were forced to sell these equity stakes, adding to the downward pressure on the stock market.

### 10.6.3   The Crisis Intensifies

By the start of 1997, it was clear that some large Japanese banks might fail. This was a solvency problem: the Bank of Japan had succeeded in avoiding a liquidity crisis by pouring funds into the market[422], but there was little action to improve the banks' solvency. Some banks succeeded in raising new capital – often assisted by the 'coordination' of the Ministry of Finance – but this was not enough to provide them with a sufficient cushion. The state was forced to intervene, injecting capital into a number of banks and, in some cases, merging institutions. A Financial Reconstruction Commission was set up to manage the bailout, and a state-owned 'bad bank', the Resolution and Collection Corporation, bought problematic loans from the banks allowing them to clean up their balance sheets.

This strategy was modestly successful. Japan's growth was not impressive, but it was at least improving by late 1999[423]. Stock prices staged a modest rally too, and the credit spread of the larger Japanese banks decreased. It had taken more than a decade, but Japan's financial system had begun to stabilise.

## 10.7   A Comparative Anatomy of Financial Crises

The Lost Decade in Japan, the Great Depression and the Credit Crunch are separated by eighty years and eight thousand miles, yet they show considerable similarities. Here we look at some of the common features that are shared by all three crises.

### 10.7.1   Underlying Causes of a Crisis

Six phenomena haunt financial crises with depressing regularity:

- An out-of-date regulatory environment;
- Low borrowing costs, with both absolutely low government rates and low credit spreads: *cheap money* for short;
- Credit concentrations;
- Funding liquidity risk and/or the assumption that currently good asset liquidity conditions will persist;
- Leverage; and
- Loose lending.

The regulatory environment is crucial because big changes are typically only possible in financial regulation after a crisis. The 1930s saw huge changes

which served the U.S. well for fifty years: similarly the post-war reconstruction of Japan created a financial system that was robust for a while. As time
passes, though, these crisis-motivated financial architectures are modified and
liberalised. Deregulation – as in the dismantling of the Glass-Steagall Act discussed in section 7.2.4 – loosens historical structures that are seen as holding
back the development of the financial system without putting compensating
controls in place. The resulting rush often endangers financial stability.

In 1980s Japan, deregulation led to pressure on credit spreads as more financial institutions were drawn into lending. Increasing competition caused
decreasing bank profitability, and some institutions were drawn into taking
larger and more complex credit risk positions than they were equipped to
handle. More competition forced banks to lend more laxly, and the combination of rising asset prices and cheap borrowing encouraged individuals and
corporates to increase their leverage. This is exactly the same behaviour as
we saw in the Greenspan Boom or in the margin lending boom of the late
1920s[424].

### 10.7.2  The Loss of Confidence

The consequence of too much lending is losses for banks when the bubble bursts. The longer the bubble lasts, the worst the excesses at the peak, and
the nastier the subsequent crisis. In fact, the term 'bubble' is slightly misleading, as it suggests that the burst will be sudden *and then all the bad news is
known.* The initial market falls can be heavy – as in the 1929 equity market
crash – but then things steadily get *worse* as market participants repeatedly increase their estimates of the losses caused by lax practices in the boom years.
Figure 10.6 illustrates the slides after the market peak in each of our three
examples[425].

In less liquid and readily observable markets such as property, the falls
are much slower. Thus for instance we have the plateau of the Case-Shiller index in 2006 (shown in figure 1.2): similarly Japanese real estate prices levelled
in 1987 and remained close to their peak for roughly five years[426]. The pattern
therefore seems to be that the prices of physical assets (particularly property)
stop rising: this eventually causes a loss of confidence which precipitates steep
stock market falls.

One reason for the pervasive loss of confidence is the lack of transparency
about losses from financial institutions. Even for fair value assets, valuation
is problematic after a crash. There is simply not enough market liquidity for
prices to be readily observable on many assets. For accrual assets, the situation
is much worse. There is a huge temptation for financial institutions not to

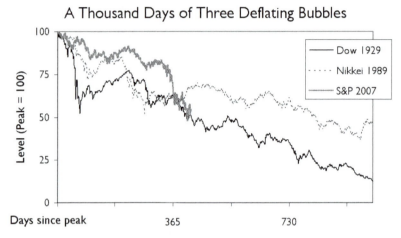

**Figure 10.6:** Equity market performance after the peak: a summary of three deflating bubbles

take sufficient write-downs or loan loss reserves in the hope that the market will recover. Sometimes this strategy works. If the market continues to fall, however, firms are perceived as lying about their asset values. Deprived of accurate information, investors fear the worst and sell financial stocks without discrimination[427].

The lack of transparency and considerable discretion of accrual account-ing – at least as it is practiced – is therefore a major factor in the loss of confi-dence in banks that characterises a crisis[428].

### 10.7.3    The Financial System Transmits Stress

The classical mechanism by which the failure of one financial institution endangers others is direct credit risk. This was evident in 1929[429]. However, as we discussed in section 10.2 earlier, there are other vectors. For the Lost Decade, a prominent one was lack of confidence in bank solvency. Steadily falling earnings throughout the economy were also important: banks simply could not generate enough earnings to recapitalise themselves fast enough. For the Credit Crunch, the run on the structured finance system and stress in the funding markets were important.

What happens after a banking crisis is crucial to the impact on the broad economy. In this situation, banks are short of capital and finding it hard to fund themselves. If they withdraw credit from the broad economy, making it much harder and more expensive for the average borrower, then any remain-ing confidence is sapped. Spending falls, non-financial companies retrench, and the recession will be long and deep. All three crises were characterised by banks restricting the supply of credit, yet Central Bankers have been slow to realise the importance of the feedback loop between perceptions of bank health and the broader economy[430]. In reality, monetary policy and financial stability are inexplicably intertwined.

### 10.7.4  Delay, then State Capitalism

It is inevitable that regulators never address a problem quickly enough. If they move too fast, then they would be accused of endangering firms. Thus they can only act once the problems are evident. That said, the authorities have sometimes delayed even longer. That was certainly true in early 1990s Japan or early 1930s America. (The case for the leading economies in early 2007 is less clear, and by late 2008, many governments had acted.) Another issue is that financial institutions may be reluctant to ask for help because that would draw attention to their problems. It is easier to help all the leading banks than one or two of them. Unfortunately procrastination sometimes exacerbates the problem, so by the time decisive action is taken – 1933, 1999 or 2008 – the bailout is bigger.

In each of the three cases, the solution was the same. Financial institutions have to be recapitalised, and the state was the only actor with the means and will to do that. Thus the RFC recapitalised banks by buying preference shares in 1930s America, the Financial Reconstruction Commission injected capital into Japanese banks in the late 1990s, and various governments are recapitalising their financial systems as I write. In each case the result is substantial state ownership of the banking system.

### 10.7.5  Rewriting the Rules

Financial crises produce a cry for more regulation. That is as much true in our time as in Roosevelt's. But perhaps there is a case for better rules rather than simply more rules. Certainly there is no shortage of volume already: the document setting out Basel 2 is thick enough to hurt if it fell on your toe. But despite this, it did not effectively constrain the leverage of large banks. The combination of capital requirements based on procyclical risk models and the ability to do old-style banking without old-style capital through the use of off balance sheet funding saw to that.

Meanwhile, the growth of fair value accounting has changed the character of banking. Diligently applied fair value can increase the transparency of bank's balance sheets. But in good times it can also provide higher earnings

| Crisis | U.S., 1929-1939 | Japan, 1989-2000 | Global, 2007+ |
|---|---|---|---|
| Key causes | Equity bought with leverage, systemic weakness | Real estate lending | Real estate lending, Funding liquidity risk |
| New risk takers | Retail stock investors | Non-bank lenders | Subprime borrowers |
| Stress transmitted by | Direct credit exposure, Bank runs | Falling asset prices, Lack of transparency | Structured finance, Funding markets |
| Confidence lost when | Stock market crashed | Non performing loan problem was hidden | Lehman went bankrupt |
| Agent of bank recapitalisation | RFC | Financial Reconstruction Commission, Deposit Insurance Corp. | TARP (US), Finance Ministries in many countries |
| Other agents of support | HOLC, Fannie, FHA, FDI | Resolution and Collection Corp. Financial Revitalization Committee | Various liquidity and asset guarantee programs |
| Notable failures or firms needing emergency support | Bank of the United States, thousands of smaller banks | Nippon Credit Bank, Sanyo Securities, Long Term Credit Bank | Bear Stearns, Northern Rock, Lehman, AIG, Fannie & Freddie |

**Figure 10.7:** Three crises compared

which, when retained, allow banks to take more risk. In bad times, though, it produces more intense deleveraging and a Credit Crunch.

Banks' behaviour was not uniquely malign or stupid. Rather they were trying to maximise returns to shareholders while obeying the rules. These rules were generous to mortgages so they made more mortgages. They permitted positions which had a lot of risk but which did not require a lot of capital, so banks took them. Increased leverage was possible through structured finance, so that became a popular tool. Overwhelmingly the rules were followed — but the rules did not serve us well.

---

## Notes

[371] For a further discussion of some of the issues here, see Stephen Ryan's *Accounting in and for the Subprime Crisis* in The Accounting Review, 2008.

[372] For a longer discussion on regulatory capital, see chapters 3 to 7 of my book *Understanding Risk*. More on regulatory capital requirements in the context of the Credit Crunch can be found in Anil Kashyap et al.'s *The Global Roots of the Current Financial Crisis and its Implications for Regulation* (2008) www.ecb.int/events/pdf/conferences/cbc5/Rajan.pdf.

[373] For a more extended discussion – albeit one that seems unaware of the sheer magnitude of the issue – see the IASB Expert Advisory Panel document *Measuring and disclosing the fair value of financial instruments in markets that are no longer active* (2008) and the companion staff summary *Using judgement to measure the fair value of financial instruments when markets are no longer active*, both available via www.iasb.org. Claudio Borio makes the interesting suggestion that firms should disclose this uncertainty, giving the readers of their accounts not just asset valuations but error bars on those estimates. See *The financial turmoil of 2007-?: a preliminary assessment and some policy considerations*, BIS Working Papers No. 251 (2007), available at www.bis.org/publ/work251.htm.

[374] Historically the first requirement was for a 'full and fair' account (the United Kingdom Joint Stock Companies Act of 1844). This was soon followed by 'full and true'. 'True and fair', the modern formulation, comes from the United Kingdom Companies Act of 1948: see Brian West's *Professionalism and Accounting* (2003).

[375] This is very much a high level summary: see www.fasb.org and www.iasb.org for more details.

[376] Strictly speaking loan loss reserves can be (but are not required to be) based on market prices, and thus the market's expectation of the value of the asset may create earning volatility depending on the methodology used to estimate these expected loss reserves.

[377] See *Wachovia Shows Why No Bank's Books Are Trusted*, Bloomberg, 30th October 2008, available at www.bloomberg.com/apps/news?pid=newsarchive&sid=aZE9yF3JayDA.

[378] See the discussion in Box 12 of the ECB's June 2008 *Financial Stability Review* available at www.ecb.int/pub/pdf/other/financialstabilityreview200806en.pdf.

[379] See the FASB's *Statement of Financial Standards No. 157* available at www.fasb.org/pdf/aop_FAS157.pdf.

[380] And indeed it would be helpful if this approach were extended to off balance sheet arrangements such as derivatives.

[381] See *Report and Recommendations Pursuant to Section 133 of the Emergency Economic Stabilization Act of 2008: Study on Mark-To-Market Accounting* available via `www.sec.gov/news/studies/2008/marktomarket123008.pdf`.

[382] The QSPV rules are in FAS 140, available at `www.fasb.org/pdf/fas140.pdf` and FIN46(R), available at `www.fasb.org/pdf/fin46R.pdf`. The door was closed in 2008 when the FASB changed the rules, but given that this was after the effective closure of much of the securitisation market, the horse could be said to have already bolted. For a further, more or less layman's account of the U.S. consolidation rules, see the Report of the CRMPG III, *Containing Systemic Risk: The Road to Reform* (2008) available at `www.crmpolicygroup.org/docs/CRMPG-III.pdf`.

[383] FASB Chairman Robert Herz gave his own perspective in a speech on the 8th December 2008, *Lessons Learned, Relearned, and Relearned Again from the Global Financial Crisis–Accounting and Beyond* available at `www.fasb.org/articles&reports/12-08-08_herz_speech.pdf`. He says: 'neither FAS 140 on transfers of financial assets nor FIN 46(R) on variable interest entities are God's gift to accounting. While the basic concepts underlying these pronouncements are relatively simple, their implementation can often be challenging… It seems that some folks used Qs like a punch bowl to get off-balance sheet treatment while spiking the punch… we are proposing significant revisions to both FAS 140 and FIN 46(R) that we believe will result in many, if not most, securitizations and VIEs being on balance sheet.'

[384] Even for banks, failures are regular. The FDIC for instance reported numbers of failing banks during much of the 1980s in the high teens or low twenties. See *Federal Deposit Insurance Corporation Changes in Number of Institutions, FDIC-Insured Commercial Banks in the United States and Other Areas* available at `www.fdic.gov`.

[385] To quote the committee: "One important objective of the Committee's work has been to close gaps in international supervisory coverage in pursuit of two basic principles: that no foreign banking establishment should escape supervision; and that supervision should be adequate." See `www.bis.org` for far, far more details.

[386] The rules are slightly more complex than this – government lending was divided into loans to OECD governments, for instance, which were 0% risk weighted, and others, which were not, but still Basel 1 is only 30 pages long. See `www.bis.org/publ/bcbsc111.htm` for details.

[387] In fact, arguably, it never was: for many banks interest rate risk in how the bank is funded was and continues to be more important than credit risk, but this risk class has been systematically ignored in regulatory capital requirements.

[388] The precise requirement under the models rules is to have capital for market risk equal to three times the ten day 99% value at risk. Since VAR scales roughly as the square root of time, that is equivalent to a 99% 90 day (i.e. one quarter) VAR. One quarter in a hundred is one year in twenty five. For more details on the market risk rules, see chapter four of my book *Understanding Risk* or Philippe Jorion's *Value at Risk: The New Benchmark for Managing Financial Risk* (2000).

[389] In reality, then, backtesting can help to invalidate a risk model which makes predictions at a one year confidence interval 95% or higher, but it can never validate it.

[390] The topic of the utility, or otherwise, of the internal models promulgated by the 1996 market risk amendment could occupy an entire book. Certainly VAR models give little information about how much you might lose in that 1% of bad outcomes. Moreover there is

some evidence that financial institution's risk taking has become less normal over time, and hence the VAR alone may be becoming a less useful measure. Nevertheless having *some* integrated risk measure is better than having none, and VAR supplemented by various other measures can be a useful tool in risk management. See Leonard Matz's *The Use and Misuse of Value-at-Risk Analysis for Bank Balance Sheets* (Bank Accounting and Finance 2005) available at `www.gloriamundi.org/picsresources/lmuse.pdf` and Jón Daníelsson's *The Emperor has no Clothes: Limits to Risk Modelling* (Journal of Banking and Finance, Volume 26, 2002) for longer discussions.

[391] The Basel Committee now only advertises the second consultative paper (of 2001) on its website: see `www.bis.org/bcbs/bcbscp2.htm`.

[392] See *International Convergence of Capital Measurement and Capital Standards*, A Revised Framework, Comprehensive Version 2006 available at `www.bis.org/publ/bcbs128.htm`.

[393] Although by no means all: securitising the mezz tranches while retaining a thin, high yield junior and a AAA-rated senior tranche can achieve a significant capital arbitrage for higher spread collateral.

[394] Technically the formula gives the incremental extra capital required at 99.9% to cover the contribution of one new exposure with the given correlation to an already diversified portfolio: see Hugh Thomas and Zhiqiang Wang's *Interpreting the internal ratings-based capital requirements in Basel II* (Journal of Banking Regulation, Volume 6, 2005) for more details.

[395] The reason for this, anecdotally, is that when the supervisors reviewed the performance of the bank's own credit risk models during the 1998 Russia/LTCM crisis, they were so perturbed by the results that they decided not to permit the banks to estimate this crucial parameter.

[396] Would it be too unreasonable to suspect that Basel 2 was designed by old fashioned bank regulators for old fashioned banks, and the devil take anyone else?

[397] See their paper *What we know, don't know and can't know about bank risk* (to appear in *The Known, The Unknown and The Unknowable in Financial Risk Management* edited by F. Diebond et al.) available at `www.newyorkfed.org/research/economists/schuermann /Kuritzkes_Schuermann_KUU_23Mar2008.pdf`.

[398] Speech, 8th October 2004, available at `www.fsa.gov.uk/Pages/Library/Communicat ion/Speeches/2004/SP201.shtml`. Despite Sharma's remarks in 2004, comprehensive reform of the UK liquidity risk regulations did not happen until 2008. The proposed new rules are discussed in FSA Consultative Paper CP 08/22 *Strengthening liquidity standards* available at `www.fsa.gov.uk/pages/Library/Policy/CP/2008/08_22.shtml`.

[399] Capital is required for market risk as measured by VAR, but not for asset liquidity risk, nor for tail risks not captured by VAR.

[400] See her testimony *Basel II Regulation: U.S. Market and Competitiveness Implications* before the Subcommittee on Financial Institutions and Consumer Credit and the Subcommittee on Domestic and International Monetary Policy, 11th May 2005, available at `financialservic es.house.gov/media/pdf/051105ksp.pdf`.

[401] For more background, see E. Catarineu-Rabell et al.'s *Procyclicality and the new Basel accord: banks' choice of loan rating system*, Economic Theory Volume 26, 2005 or Michael Gordy and Bradley Howells's *Procyclicality in Basel II: Can We Treat the Disease Without Killing the Patient?* 2005, available at `www.bis.org/bcbs/events/rtf04gordy_howells.pdf`.

[402] For details see *Results of the fifth quantitative impact study*, Basel Committee on Banking Supervision, 2006 available at `www.bis.org/bcbs/qis/qis5results.pdf`. The most

damning data is in table 11, which shows that the largest saving of any product line under the IRB in Basel 2 is retail mortgage lending.

[403] See section 3 of *The Run on the Rock*, House of Commons Treasury Committee 2008 available via `www.publications.parliament.uk/pa/cm200708/cmselect/cmtreasy/56/5602.htm`.

[404] For a further discussion of some of the issues here, see Patrick Honohan's *Risk Management and the Costs of the Banking Crisis*, Institute for International Integration Studies (2008) available via `www.tcd.ie/iiis/documents/discussion/pdfs/iiisdp262.pdf`.

[405] Paul Kupiec was one of the first to point out some of the model risks in Basel 2. His papers *Basel II: A Case for Recalibration* (available at `www.fdic.gov/bank/analytical/cfr/2006/wp2006/CFRWP_2006_13_kupiec.pdf`) and *Financial Stability and Basel II* (Annals of Finance, Volume 3, 2007) discuss some of the issues surrounding the specification of default correlation used in the New Accord.

[406] Switzerland has taken an interesting step in this direction. From 2013 its two largest banks will have to comply with a simple leverage ratio requirement as well as stricter capital rules. See a Reuters story by Lisa Jucca, *Swiss up deposit insurance, set tough capital rules*, 5th November 2008, available at `www.reuters.com/article/GCA-CreditCrisis/idUSTRE4A48J020081105`.

[407] The estimation of improbable events is plagued with model risk. The less likely they are, the more robust the model has to be for model error not to swamp the original estimate. See Toby Ord et al.'s *Probing the Improbable: Methodological Challenges for Risks with Low Probabilities and High Stakes* (2008) available via `arxiv.org/abs/0810.5515`.

[408] This is probably a mistake. While it makes sense for, say, Deutsche Bank to be supervised under the same rules as Citigroup or BNP Paribas, it does not seem necessary to use the same supervisory framework for small specialist firms as for universal banks.

[409] The differences between the SEC regulatory regime – which applied to Bear Stearns, Goldman Sachs, Lehman Brothers, Merrill Lynch, and Morgan Stanley – and the Basel 2 one are considerable. It seems that the SEC rules permitted significantly higher leverage than Basel 2. Sadly a direct comparison is difficult. One might have hoped that the acquisition of Bear Stearns' portfolio by JP Morgan would have provided a useful data point: JP's capital requirements before and after the purchase would have given the Basel 2 requirements for Bear Stearns' portfolio, and this could have been compared with the Bear Stearns' reported capital under the SEC regime. However the FED granted JP Morgan an exception from the need to calculate capital for this portfolio (see `www.federalreserve.gov/BoardDocs/LegalInt/FederalReserveAct/2008/20080403/20080403.pdf`). A thorough analysis of just how much more generous the SEC's rules were would be useful in any post-mortem on the broker/dealers.

[410] For some steps in this direction, see the ECB's *Credit risk transfer by EU banks: activities, risks and risk management* (May 2004) available at `www.ecb.int/pub/pdf/other/creditrisktransfer200405en.pdf`.

[411] For a broader comparison of insurance vs. banking in this context see *Risk Management Practices and Regulatory Capital – Cross Sectoral Comparison*, Basel Committee on Banking Supervision, 2001 available at `www.bis.org/publ/joint04.pdf`. This paper also makes the reasonable point that comparing capital requirements alone can be misleading due to the different definitions of eligible capital between banks and insurers, and the relatively large rôle of reserves for insurers. Notice however that these differences should not obscure the *comparative advantage* of different asset classes. If insurers have a comparative advantage in

taking (unfunded) credit risk – as I believe they did in the Boom years – then credit risk will leave the banking system for the insurance system.

[412] See Table 3 in the section 'Standard & Poor's Capital Adequacy Model' from Standard and Poor's *Global Bond Insurance 2007 Book*, available via `www.standardandpoors.com`.

[413] Basel 2 sets a minimum probability of default in some cases of 0.03%. This is far too low: a figure at least ten times higher would be more appropriate.

[414] Of course we are assuming that the accrual accounter is able to accurately estimate the required LLR for the long term default costs of the assets. This is questionable.

[415] Notice that the problem is not with fair value: it is that all retained earnings count as (Tier 1) capital. One way around the problem would be to split retained earnings into realised earnings and unrealised, with a restriction as to the fraction of capital that could come from unrealised retained earnings.

[416] The equity of an institution whose risk is dominated by fair value accounted credit exposure will therefore act like a call on the market. This observation helps to explain the meteoric risk of Countrywide Financial's share price and its sudden fall.

[417] See Akihiro Kanaya and David Woo's *The Japanese banking crisis of the 1990s: sources and lessons*, IMF working paper WP/00/7 (2000) available at `www.imf.org/external/pubs/ft/wp/2000/wp0007.pdf`.

[418] For more on the Lost Decade, see Kazuo Ueda's *The Japanese banking crisis in the 1990s*, BIS policy papers (2005) available at `www.bis.org/publ/plcy07q.pdf` or, at a higher level, Hiroshi Nakaso's *The financial crisis in Japan during the 1990s: how the Bank of Japan responded and the lessons learnt*, BIS Papers No. 6 (2001) available at `www.bis.org/publ/bppdf/bispap06.pdf`.

[419] The macroeconomic and monetary policy aspects of the Lost Decade are dealt with extensively in the other references so we do not go into a lot of detail here. Suffice it to say, when rates were close to zero, the Bank of Japan had nowhere to go: it was caught in a *liquidity trap* whereby even zero rates did not provide sufficient stimulus. A solvency-challenged banking can consume cash as fast at the Central Bank is willing to print it.

[420] Two credit cooperatives, Tokyo Kyowa and Anzen, were amongst the first to fail. See Nakaso's paper for a more detailed chronology.

[421] There have however been some troubling trends in the application of fair value which definitely increase the number of hiding places. See for instance the IASB's *Amendments permit reclassification of financial instruments* (October 2008) available at `www.iasb.org/News/Press +Releases/ IASB +amendments +permit +reclassification +of +financial + instruments.htm` and the SEC's *Disclosure Regarding the Application of SFAS 157 (Fair Value Measurements)* (March 2008) available at `www.sec.gov/divisions/corpfin/guidance/fairvalueltr0308.htm`. Admittedly the IASB acted after pressure from the European Commission: see `ec.europa.eu/internal_market/accounting/docs/letter-iasb-ias39_en.pdf` but still this new opacity is to be regretted.

[422] For a longer description of monetary policy in Japan and other crises, see Paul Krugman's *The Return of Depression Economics and the Crisis of 2008*.

[423] Although it took much longer for economic growth to consolidate in positive territory: see the Ministry of Internal Affairs and Communications *Statistics Handbook Japan* available via `www.stat.go.jp/english/data/handbook/index.htm` for the data.

[424] As Paul Krugman points out, Japan also suffered from a lack of clarity about what was and what was not government guaranteed. Much like Fannie and Freddie, there was a broad belief that the large banks would be supported, but this was not explicit.

[425] For more on the aftermath of crashes, see Carmen Reinhart and Kenneth Rogoff's *The Aftermath of Financial Crises* (2008) available at www.economics.harvard.edu/faculty/rogoff/files/Aftermath.pdf.

[426] See Figure 2 in Kiyohiko Nishimura and Yuko Kawamoto's *Why Does the Problem Persist? "Rational Rigidity" and the Plight of Japanese Banks*, RIETI Discussion Paper Series 02-E-003 (2002) available at www.rieti.go.jp/jp/publications/dp/02e003.pdf.

[427] A comment from Meredith Whitney writing in the Financial Times (on the 30th November 2008) is worth repeating here: 'there is more transparency in off-balance-sheet master trust data than in on-balance-sheet accrual accounting.' What she means is that at least in a securitisation, investors will not allow the issuing vehicle (which in a typical U.S. credit card deal is a master trust) to smooth its earnings. They want to see the truth as the only thing supporting their investment is the performance of the underlying assets. Investors in bank equity are not as demanding.

[428] Jonathan Weil's broader point about Wachovia's accounting (discussed earlier) is that 'nobody with any authority is calling them on it'. Certainly cases like this one make it hard to have confidence that accrual accounting as practiced really does always give a true and fair value of a company.

[429] To be fair, systemic weaknesses such as the absence of universal deposit protection – making banks highly vulnerable to deposit runs – and the predominance of small, badly capitalised banks were also important in the Great Depression.

[430] As Nakaso, referred to previously, discreetly puts it: 'the Bank [of Japan] did not necessarily have a clear vision of how this was going to feed back on the real economy.'

# Chapter 11

---

## Changes and Consequences

*Of man's first disobedience, and the fruit*
*Of that forbidden tree, whose mortal taste*
*Brought death into the world, and all our woe,*
*With loss of Eden.*
John Milton

---

## Introduction

Ordinarily, the doings of the financial system are fascinating to some, and –
let's be honest – deeply boring to others. But when banks stop lending, everyone is affected. No major economy can prosper without credit. Growth
falls, unemployment rises, pension funds lose value. Without vigorous government action, a deep recession is likely.

We begin by looking at how the banks' problems in the early stage of
the Crunch become problems for everyone later on. Once the mechanism by
which problems are transmitted is clear, we can discuss some of the policy
alternatives available not just for alleviating the Crunch, but also for longer
term reform of the financial system. A wide debate about these changes is
important as we will be bearing the cost and consequences of them for many
years.

## 11.1   Transmission

A Credit Crunch occurs when there is a widespread contraction in the availability of credit. The nature of the financial system means that this is accompanied by an increase in credit spreads, an increase in risk aversion, and a
flight to high quality liquid assets by investors.

Credit is provided by the banks and the securities markets. Why did both
of these mechanisms fail at once?

### 11.1.1  Two Spirals

There are two spirals which together explain this failure. The first one relates to solvency:

- Bad mortgages are made and do not perform. The holders of mortgage risk take losses. This affects financial institutions in two ways: through retained mortgages; and through purchased MBS.

- Assets linked to MBS fall too: this includes ABCP, since conduits and SIVs held MBS; and CDOs of ABS containing MBS. Faced with plunging prices on some AAA-rated securities, investors start to distrust many structured investments.

- Margin calls and the forced liquidation of some leveraged investors cause further price falls.

- A significant amount of mortgage risk is taken on an unfunded basis. These risk takers suffer significant losses, and this in turn decreases the quality of all of the protection they provide.

- These phenomena create a run on the structured finance system as investors withdraw.

- Investors, including financial institutions, lose confidence in their risk assessment and withdraw risk capital. Risky asset prices fall leading to further losses and a flight to quality.

- Illiquidity causes asset price uncertainty. This further decreases confidence about the credit quality of financial institutions. Opaque financial statements and inadequate disclosures exacerbate the problem.

- Losses erode capital, reducing risk taking capacity, including lending capacity.

- Procyclical capital requirements and risk models also cause reductions in risk capacity.

- If asset values fall far enough, even sound banks become unsound.

The second spiral relates to liquidity. It interacts with the solvency spiral in several places.

- A lack of confidence in financial institutions reduces their willingness to lend to each other. Other investors withdraw their funding too.

- Collateral calls cause increased demand for funding, and the demand for good quality collateral intensifies the flight to liquidity.

- A higher cost of funds decreases financial institutions' profitability and further restricts their ability to lend.

- The collapse of the structured finance system increases the demand for funding as assets come back on balance sheet.

- Increased stress in the funding market and the desire to communicate their soundness to the market causes financial institutions to reduce their funding liquidity risk.

- Central Bank intervention may not be sufficient to reliquify the financial system if the loss of confidence is too severe.

These two interacting spirals cause a dramatic reduction in lending[431]. Deprived of sufficient credit, the broad economy contracts. If the contraction is too severe, a recession is inevitable, even if the initial cause is relatively minor[432]. How can the landing be made a little less bumpy?

### 11.1.2    Reducing the Transmissions

The first key tool is to reduce the likelihood of problems in the first place. This involves modifications to the mortgage market and changes in how the financial system provides credit to the economy. It also requires firms to improve their risk management practices, culture, and internal incentive structures.

The second step is to reduce the *coupling* in the financial system: that is, to impose measures to stop a local problem turning into a global problem. These measures would make it harder to slip down the spirals. They involve (to return to the terminology of the Introduction) rewriting the rules of the system – both accounting and regulatory rules – to make each firm less dependent on the performance of the others and on the state of the market. These two steps are the topics of the next two sections.

## 11.2    The Provision of Credit to the Broad Economy

With 20/20 hindsight, the way that the financial system provided mortgage credit was flawed. To begin with, mortgage structures were sub-optimal for both borrowers and lenders, setting up conditions so that a housing bubble was inevitable.

The bursting bubble causes losses. But then the system was insufficiently robust, and relatively minor mortgage losses led to a major recession. First, mortgages. How can this market be less susceptible to boom and bust?

### 11.2.1  Modifying Mortgages

The house price falls we discussed at the start of the book were not too severe. But the U.S. housing market got worse through the whole of 2008. By the end of the year, for instance, prices in Southern California had fallen 47% from their peak, and foreclosed homes accounted for 56% of all sales[433]. Foreclosures ruin lives as well as banks. Clearly it would be good to reduce the volume of mortgage defaults. In the long term it is not sensible to base economic prosperity on rising property prices, but in the short term, helping mortgage borrowers is important.

Rates had fallen so far during 2008 that most borrowers who could re-finance, had. But in order to be able to refinance, an acceptable LTV was needed. And that was a problem for many loans, especially those originated between 2005 and 2007, due to the fall in house prices.

The obvious answer in the short term is to encourage mortgage servicers to modify loans, refinancing good borrowers into level pay loans with lower principal balances. But often this is not possible: the legal agreements which were made when the mortgages were sold to the securitisation SPV often limit the ability to modify the loans. In some cases this is true even when such a modification would be ultimately in the interest of the tranche holders[434]. Clearly in retrospect giving responsibility for modification to someone without an economic interest in the results of that activity – the servicer – then limiting what they can do is not the best arrangement for either investors or mortgage borrowers.

Another difficulty arises due to American bankruptcy law. Currently this prohibits judges from altering the terms of mortgages for primary residences, making it much harder for distressed borrowers to restructure their debts. There is a proposal that would give bankruptcy judges greater freedom to modify loans, but mortgage lenders are lobbying hard against this[435]. Providing a route out of financial distress which allows borrowers to keep their homes could reduce both eventual losses and the damage to society. But it would involve banks recognising losses earlier, something that they are currently loath to do.

There are moral hazard arguments against providing state support to people who opportunistically took out mortgages as a bet on house prices, knowing that they could not afford their loan in the long term. Perhaps own–to–rent arrangements, whereby some of these borrowers lose their right to any eventual house price recovery, are sensible here. But many borrowers were mis-sold mortgages. Foreclosing on these victims of financial innovation as their teaser rates end is not just inequitable: it also risks intensifying the fall in house prices. This problem is significant given that some estimates suggest that nearly half of recently originated subprime mortgages were extended to borrowers that would have qualified for GSE-conforming loans[436]. If the U.S. government can spare $700B of TARP funds to help financial institutions, perhaps it can also find some funds for ordinary Americans who are trying to pay their mortgages but risk losing their homes when low rates are no longer available thanks to their loan structure[437]?

### 11.2.2   Mortgage Structures and Walking Away

It is time to consider restricting the range of mortgages available to those that encourage good behaviour on the part of borrower *and* the lender. One possibility would be to forbid loans with an LTV greater than 80% or a PTI above 38%[438] based on proven income and the maximum loan rate. The legislative effort needed to make recourse loans the standard structure in all states would be considerable, but there would be many benefits if that could be done too. Confidence in mortgage lending will only return when mortgage lenders know that they are not taking a massive bet on the future level of house prices.

Another reform here would be to eliminate teaser rate loans, and standardise on level pay structures[439]. The Danish mortgage market may offer a useful model: here not only is the standard structure a prepayable, fixed rate amortising recourse loan with an LTV less than 80%, but there are also strict matching rules governing how every mortgage is funded. These pass prepayment risk on to mortgage bond investors, ensuring that banks take minimal funding liquidity risk. Standardisation on low risk loan structures (and other features of the Danish market[440]) means that mortgage credit is always available at a uniform rate. This model does make it harder to qualify for a mortgage, but it also makes it considerably less likely that mortgage lending will endanger the banking system.

### 11.2.3   Funding the Economy: Forcing the Horse to Drink

Society allows banks to operate because it judges that they perform a useful function without excessive cost. If that judgement changes, then the unique bargain of banking makes less sense. Why should financial institutions be allowed to use deposits for their benefit rather than that of the depositor? The only justification is that the financial system provides credit to the broad economy. But by the third phase of the Crunch, this credit provision was stalling. Firms were deleveraging, raising capital (mostly from the state) and reducing their assets. They were not lending, despite various government schemes[441].

Perhaps one immediate solution is to make access to the central bank window dependent on new lending. That is, as part of any access to central bank funding, financial institutions must pledge a certain percentage of newly created credit. This would strongly encourage them to lend, and place a premium on recent loans.

### 11.2.4   Government Bailouts

The roll call of financial institutions that have gone bankrupt, had to sell themselves, or been nationalised during the Credit Crunch is long. Banks like Washington Mutual, Northern Rock and Wachovia; retail mortgage specialists like Countrywide, Fannie and Freddie; broker/dealers like Bear Stearns, Lehman Brothers and Merrill Lynch: all disappeared. More or less every other large financial institution has benefited from some measure of state support including liquidity injections, recapitalisations, and asset guarantees. The diversity and size of government interventions is dizzying[442].

These measures were not enough, perhaps because they were incremental, with no single measure of sufficient size to arrest the slide[443]. By the end of 2008, the Credit Crunch was more intense than ever. Even Citigroup – once the largest bank in the world – was under threat and breaking itself up[444]. If the steadily growing stream of taxpayer's money hadn't worked, what would?

At this point it was natural to ask if widespread bank nationalisation was necessary. Should the government buy the banks, clean up their balance sheets, and sell the resulting cleaned up institutions back to the private sector? This strategy is sometimes known as the Swedish model after the approach taken to that country's banking crisis in the early 1990s[445]. It was successful in the past: perhaps, given time and money[446], it would work again.

It is interesting that in the current debate even more radical alternatives have not been widely discussed[447]. For instance, should the government not just take but keep control of the commanding heights of the economy[448]? Is it time to consider the *narrow banking* model whereby deposit taking institutions can only use those deposits to buy government bonds? There are two key questions here:

- *What is the right balance between financial stability and the cost of credit?* We have seen that we can have mostly cheap credit at the price of occasional dramatic financial crises. We can clearly have a stable banking system if leverage is much lower and banks' ability to take risk is dramatically constrained. But that will make credit more expensive and constrain growth.

- Many economists think that financial crises are an inevitable part of market capitalism[449]. If we accept this, *should the state take a permanent stake in the financial system?* This stake may be larger in a crisis and smaller in more normal times, but if we have strong reason to suspect that state funding will be needed on a regular basis, perhaps the taxpayer should have more benefit in the good times too.

## 11.3   What Worked and What Didn't

In 1988 the USS Vincennes, a warship, shot down a passenger airliner. 290 civilians were killed. How did such a tragic mistake happen?

The plane was shot down because the Vincennes thought that it was a hostile military fighter. The aircraft involved – Iran Air 655 – was an Airbus. The Vincennes thought that it was an F14. The warship fired despite the fact that the plane was following its designated flight plan and flagging its presence using commercial standard radio transponders[450]. Disaster might even been averted if someone had looked out of a window. But the Vincennes' Combat Information Centre did not have any windows; instead, the ship's commanding officers relied exclusively on computerised information systems to provide their risk information.

The Vincennes was in the middle of skirmishes with small boats at the time of the incident. Perhaps an air attack was expected: certainly the ship's officers had trained for one. Neutralising a hostile aircraft was part of their job description. In contrast they were not trained to operate in a situation where civilian aircraft had to be distinguished from potential threats[451]. The Vincennes did not even have the right equipment to allow them to monitor civilian air traffic control frequencies.

In short, then, the Vincennes had a military model of the world. The key decisions were taken based on the assumption that this model was correct. If anyone thought that there was an alternative explanation for the information they had, or that this information might be wrong, it did not stop them deploying lethal force.

The shooting down of Iran Air 655 is instructive in that it illustrates that confident but erroneous risk assessment is not confined to finance. Banks did not acquire tens of billions of dollars of subprime-backed securities thinking that they were risky. They bought them confident that they weren't. No one died as a consequence of the banks' errors, but a trillion dollars of losses and a faltering global economy are nevertheless serious. What then do we learn from the many mistakes that were made in risk assessment?

### 11.3.1 'I told you so'

Sometimes people ask 'Was Bear Stearns (or Northern Rock or Lehman or Fannie or Freddie or AIG or ...) stupid or greedy?' It is an obvious question, but it might not be helpful to suppose that it has an answer. First, the senior management of large firms have such a diversity of opinions that it hardly makes sense to attribute something like stupidity to a corporation. Most decisions in large firms are debated extensively. There is almost always someone who claims that any particular action will lead to disaster. Mostly, it doesn't. So the charge that firms did not listen to internal whistleblowers is vapid: if the Cassandras always prevailed, firms would never do anything. A better question, then, is: 'why didn't many firms' controls prevent disaster?' Or, as Alan Greenspan put it, why didn't the 'self-interest of lending institutions to protect shareholders' equity'[452] lead to better risk management?

### 11.3.2 Understanding Valuation, Understanding Earnings

Financial risk is the risk of a loss. But to say that you have lost money implies that you know what the value of your portfolio is in the first place. Valuation, then, is the foundation of risk management. Without accurate valuations, and a good understanding of the uncertainty in them, further measures are of limited use.

This is especially the case as the confidence of investors depends on their beliefs about firms' financial reporting. If they suspect that they are being lied to, and that a firm is not really worth what it claims, then they will be very reluctant to lend to the firm, and it will suffer a funding crisis. Therefore the ability not just to value your portfolio properly, but also to convince investors that you are valuing your portfolio properly, is key to mitigating funding liquidity risk. This is one reason that banks with significant accrual accounted assets were suffering in late 2008: investors are simply not confident that they are solvent.

Some firms did not understand these issues. They did not have robust controls on valuations, nor sufficient understanding of valuation uncertainty.

They thought that just because investors had been happy with their disclosures in the Boom years, these scant details would suffice in the Crunch. Perhaps they did not understand how confidence could be sapped by concerns about the ability to smooth earnings in accrual accounting. They were too slow to discuss the detail of their valuation processes, and too confident of their bluster. This would not matter so much were the institutions concerned the only victims: but widespread inadequacies here sapped confidence in the financial system as a whole. In the future, many institutions will have to be more robust in their valuation processes; they will have to disclose much more about how valuations and loan loss provisions are determined; and they will have to be more open about valuation uncertainty.

### 11.3.3    Lessons in Risk Management

Senior management sometimes tend to assume that the risk control process in their firm works. But there are an awful lot of things that are necessary for this to be true. These include[453]:

- The information going into the risk management process is complete and correct;

- The process does the right thing with the information quickly and reliably;

- The resulting risk information is distributed to key individuals;

- Those individuals act on the information; and

- The process is revised as needed given the firm's changing balance of risks.

The risk control process is rather complicated so it can fail in a number of ways. Some common pitfalls are:

- *Inappropriate technology.* In many cases firms had risk aggregation systems which either did not capture some risks or did not measure them accurately.

- *Simplifying assumptions.* Financial risk is enormously complicated. There are many thousands of risk factors, and they interact in complex ways. It is impossible for any large firm to manage their risk without some simplifying assumptions. But once these assumptions are built into risk systems, they are hard to question, despite the evolution of the markets and of risk taking.

- *Culture and Silo management.* It can be difficult to be openly sceptical of business groups who seem to be making hundreds of millions of dollars. The more golden eggs the geese lay, the harder it is to question where the eggs come from. In a Boom, then, there is a natural tendency for risk management to lose political power within a firm to the money makers. Moreover, individual managers, desirous of their own fiefdom within the firm, discourage the flow of information from their groups. There can be little sense of broad responsibility for the firm's position, and little open discussion of risk.

- *Incentive structures.* Often, risk control staff are seen as holding back business. Those who can claim that they have made the firm money are well rewarded, while those that simply stop the firm from losing money make much less, and have less authority.

- *Failure to manage.* The complexity of risk taking, particularly in structured finance, means that managers sometimes do not understand the risks that they are responsible for. And if all seems to be going well, why concern yourself with the details?

None of these pitfalls are particularly surprising: as a consequence of their ubiquity, more or less every firm fails to measure all of its risks accurately. Moreover even those firms that do accurately measure the important risks sometimes fail to act on those measurements. It is easier to produce risk reports showing a growing problem than to reverse the trend.

The Crunch is strengthening risk management within many firms. In the future, financial institutions will have better systems, better reporting, and more controls. But they will also need scepticism: good risk management in part depends on continually doubting whether your risk management is good. And mathematics cannot substitute for simple controls, like looking out of the window, or not buying tens of billions of dollars of the same type of risky security.

### 11.3.4    The Tail of Risk

Much of risk management is about *the tails*, that is unlikely events. Successful here requires being rather creative about your notion of what can happen. Thus stress tests – which investigate the impact of large changes in the market or other unlikely events – have an important rôle in risk management[454].

This is especially so as there is a powerful incentive for firms to take more tail risk, that is to take bigger and bigger bets that these unlikely events will not happen. Here's why. If you set a threshold for measuring risk, then you encourage risk taking beyond the threshold. Thus saying that the firm will measure risk at 99% confidence means that traders will bet on one in a thousand events as these are invisible to the firm's risk system[455].

### 11.3.5 Market Risk Misjudged

The problem of managing tail risk taking is most evident in market risk modelling. Here most firms use a type of internal model permitted in the 1996 Market Risk Amendment (discussed in section 10.2.3). These models were originally designed for relatively simple types of position[456]. Over the years, risk taking had got more sophisticated, but the risk models at all firms had not kept pace.

To see the size of the issue, consider a firm with a good reputation for risk measurement: JP Morgan. In 2007, this bank's internal model estimated that the amount the firm could lose on a one in a hundred day event was roughly $100M. But the firm chose to keep $9.5B of capital against market risk, far more than the regulatory rules required[457]. This indicates the gap between the capital an internal model with a 99% confidence interval estimates is necessary for safety and the amount a prudent firm actually keeps. As the chairman of the FSA said,[458] "We need to ... increase capital requirements not just marginally but several times. The present system of capital regulation of trading books is from a prudential point of view seriously deficient." Certainly those firms that were not as prudent as JPMorgan – perhaps because they trusted their internal models more – found themselves suffering losses which were much bigger than the capital supporting them. Prudence was a necessary survival trait in the Crunch.

### 11.3.6 Unfunded Risk Taking

Counterparty risk assessment is a basic discipline for financial institutions. If someone might owe you money in the future, you want to know if they are going to pay you. Clearly this control failed: how else can we explain the banks buying billions of dollars of protection from the monolines? After all, these firms were unlikely to be able to afford to pay out if the banks needed to make a claim.

Unfunded risk taking makes the financial system more efficient. But it only has a future if financial institutions can be more effective at assessing and managing counterparty risk. Higher collateral requirements from day one will be important here, rather than the dangerous situation whereby collateral is required only after a downgrade[459]. AIG's problems are evidence enough of the dangers here.

### 11.3.7 Liquidity Risks

Another clear failure in the risk management of some financial institutions has been their handling of funding liquidity risk. Perhaps because the consequences of too much risk in this area are felt only infrequently, but the benefits are available most of the time, many firms were reckless here. In reality, funding liquidity risk and solvency risks are equal, interacting partners. As Charles Goodhart said, "Liquidity and solvency are the heavenly twins of banking, frequently indistinguishable. An illiquid bank can rapidly become insolvent, and an insolvent bank illiquid"[460].

Another key fact made plain by the events of the Crunch is the dependence of asset liquidity risk on funding liquidity risk. If other people get their funding liquidity risk management wrong, then they are forced to sell assets, and if you hold those or related assets too, your portfolio may become illiquid thanks to their sales. Note in this context that short term repo is only an effective funding tool if the asset being funded remains liquid. Therefore it should only be used as a small percentage of a firm's funding mix.

Asset liquidity can change fast. Some firms made implicit liquidity assumptions both in their business strategies[461] and in their risk management techniques. In the future these will need to be made explicit, and thoroughly questioned. All firms will have to be able to prove that they are robust under dramatic changes in both asset and funding liquidity[462].

More conservative funding means higher costs. These need to be passed on within firms, so that risk takers bear these costs directly[463]. It also means that shareholders will have to get used to having less net interest income coming from term structure mismatch in funding. Perhaps they might even deign to inquire how much risk firms are taking here.

### 11.3.8 Structured Finance Risks

One lesson from the Crunch is that it is quite important to understand an asset before buying millions – or billions – of dollars of it. The principle of buyer beware was lost in the Hunt for Yield of the Boom years, and it is to be hoped that it will not be abandoned so readily in future.

To be fair, some investors found it hard to get good information about securitised assets. The asymmetric information problem whereby ABS sellers may know more about the assets than the buyers can be ameliorated by requiring that originators keep a significant amount of risk, so that at least interests are aligned[464]. And a healthy scepticism about quality of securitised assets should be fostered by all parties involved with structured finance.

### 11.3.9 Lessons for Financial Risk Management

The Crunch has exposed many risk management failures. The first step to recovery is to admit this. Despite multi-million dollar systems, and in some cases multi-million dollar bonuses for risk managers, firms did not protect themselves. The lure of short term profit above long term shareholder value was too strong. There are many detailed improvements that need to be made to firms' risk management culture, techniques and controls. But over-arching all of this must be a demand from the owners of firms that decision making is focused on adding long term value. Given that in a number of prominent cases at the moment the firms' owners include the government, the requirement to improve standards is hopefully unavoidable.

### 11.3.10 Personal Incentives

One good way of changing behaviour is to change which outcomes you compensate. Therefore any scheme which seeks to focus on long term rather than short term value creation should consider compensation schemes which are based on long term realised performance[465]. The traditional annual bonus based on (often unrealised) earnings does not provide good incentives.

## 11.4 Central Banks, Regulators and Accountants

Firms are *legally required* to be greedy. Their first duty is to shareholders. (If you hold that this behaviour is fundamentally bad for the economy, then logically you should call for a radical restructuring of Western capitalism.)

It is only possible for serious people to question what the financial system should do for society in a crisis. Certainly a minimum requirement would seem to be that it does not implode with sufficient ferocity to cause a depression. To meet these requirements we need to rewrite the rules of the games so that firms' greed does not lead to catastrophe. And at least the Crunch provides an opportunity to make these changes.

There are a variety of tools available to us. The rules determine which strategies institutions can take, and thus what behaviours the financial system can display[466]. The principal constraints on the system are:

1.   Regulatory capital requirements;

2.  Monetary policy;

3.  The cost and availability of capital for financial institutions, including state recapitalisation initiatives;

4.  The regulation of particular financial products and/or direct intervention by government bodies in the market (such as the rôle of Fannie and Freddie in the mortgage market or the guarantee of some bank assets by the FED);

5.  The nature of information providers in general and ratings agencies in particular; and

6.  Accounting standards and other disclosure requirements.

A subsection below is devoted to each of these tools. But first, what are the general principles of reform?

One way of reducing the probability and intensity of self-reinforcing spirals of asset price increases, speculation, and liquidity – which inevitably end suddenly and disastrously – is to increase the buffers in the financial system. In any situation where firms rely on each other directly or indirectly, the assumption should be that failures are possible. Counterparties can fail; liquidity can disappear; behaviours can change. Thus reform will include higher capital requirements, constraints on leverage (including minimum margin requirements), and the assumption that no risk transfer mechanism is ever completely efficient. There may also be a place for transaction taxes.

The new rules we create should apply to all systemically risky institutions be they banks, insurance companies, funds[467], or other types of firm. Simple rules are fine, as a backstop to more sophisticated model-based approaches: their rôle is to stop innovation from increasing effective leverage too far.

Confidence is a precious commodity in finance. Indeed, there is a sense in which banking is nothing but a confidence trick[468]. Therefore the new framework needs to include measures to enhance confidence. Better disclosure has a place here, as do intermediaries who can really be trusted to provide independent, expert advice.

The final principle is to accept that the state will always intervene in the market[469], and so to craft these manipulations to achieve effects that are politically desirable. Of course there is a considerable diversity of opinion on what the financial system should do for society. But by asking that question first, and crafting the framework of intervention to meet the resulting requirements, we are more likely to achieve the desired result than by beginning with a debate on what exactly should be done.

### 11.4.1   The Failures and Futures of Regulatory Capital

Regulatory capital rules are designed to provide a sufficient buffer of safety so that the failure of a regulated institution is rare and when it occurs, it is orderly. The current rules plainly do not work[470]. Whether we consider the SEC's consolidated supervised entity regime (discussed in section 7.3.1) or the Basel 2 capital regime for banks (discussed in section 10.2), leverage has not been constrained. Nor was either funding liquidity risk or asset liquidity risk. The charge sheet against the pre-Crunch regulatory capital rules is long and damning. In particular:

- The rules permitted unrealised retained earnings to be counted as capital, and hence they were used to support more risk taking;

- The rules were too complex and too easily arbitraged;

- They were procyclical;

- They placed too much faith in models, both firms' internal models and the supervisor specified model for credit risk in Basel 2;

- Specifically, they did not properly account for the growing amount of tail risk being taken by financial institutions;

- The rules encouraged the growth of 'too big to fail' institutions at the expense of smaller firms;

- They encouraged the growth of retail mortgage lending;

- Whilst giving an incentive for credit risk to leave the banking system; and

- They required no capital for important risk classes such as funding liquidity risk and interest rate risk in the banking book.

A thorough overhaul of regulatory capital is therefore needed. This must pay less regard to inherently flawed risk models, and more to overall financial stability. Regulatory capital requirements should be anti-cyclical; they should not encourage the growth of behemoth banks[471]; and they should effectively constrain both solvency[472] and liquidity risks.

The U.S. regulatory system in particular requires modernisation[473]. The patchwork quilt of multiple regulators, each with their own distinct rules, and each zealously protecting their firms, has not served society well. A single supervisor who can act as both a prudential regulator, lender of last resort, and overall guardian of financial stability, is one answer[474]. Better cooperation and some measure of rationalisation is another. But at very least the alphabet soup of the U.S. Treasury, FED, SEC, FDIC, OTS, FHFA, OCC, CFTC, and the State Insurance Commissioners needs the attentions of a hungry reform-minded government.

Another issue is the expertise of supervisors. It can be difficult for regulators to acquire staff who really understand the industry. One solution might be compulsory secondments, with every firm being required to contribute skilled, experienced staff to the supervisor. If this knowledge is combined with effective, sceptical oversight – so that regulators are not captured by the industry[475] – then perhaps the next cycle of stability-threatening innovation will be spotted in time.

### 11.4.2   *Central Bank Mandates*

If a Central Bank is concerned with financial stability, it must be concerned with asset prices. This is not just because they may give useful information on the formation of bubbles: it is also because asset prices are visible to everyone. They act as a mechanism coupling firms together. A change in prices causes fair value accounters everywhere to remark their books. Therefore Central Banks need to consider asset prices and risk taking when forming monetary policy[476].

What data might they consider? Some factors which may assist in bubble spotting include:

- *Credit spreads*, including both the default-related and non default-related components thereof. When spreads are historically low, and especially when funding liquidity premiums are low, trouble may be ahead;

- *Asset return volatility*. Again, historically low levels suggest more volatile conditions may occur in the near future;

- *The availability, amount, and cost of leverage*. Credit allows investors to turn perceptions about the value of assets into speculations, and thus it is a vital mechanism in bubble formation[477];

- *The presence of credit concentrations, and of large movements of risk, compared to historical patterns*. Anyone interested in financial stability should understand why risk is being held where it is.

- *Affordability ratios*, such as stock price/earnings ratios. When it is expensive to buy a unit of earnings, markets may be over-valued.

A key feature in bubble formation is lax monetary policy. If factors like the ones above are monitored, it may make that mistake less likely in the future.

They may also be useful in setting the anti-cyclical regulatory capital requirements discussed in the previous section.

Another aspect of Central Bank intervention concerns the use of monetary policy tools. After the Crunch, it makes sense to consider the four tools – the short rate, the money supply, eligible collateral at the window, and who can borrow at the window – as of more or less equal importance. In particular, Central Banks can tune the economy via collateral policy as well as via setting the short rate. And they will need to do this: at some point it will be necessary to wean the banking system off the current 'any collateral goes' policy. This should help to restore a more normal level of liquidity premiums if it is done correctly[478].

### 11.4.3   Rescues and Moral Hazards

The failure of Lehman showed that it can be a really bad idea to let large, systemically important financial institutions go bankrupt[479]. A new regime for distressed financial institutions is therefore required which allows early, secret intervention into troubled firms. The standard here should not be insolvency or inability to meet obligations, but something much tougher, so that a firm can be saved before it is on the brink of ruin, and before a loss of confidence fatally damages its funding.

Equity holders take risk. That means that in the event of state intervention being judged necessary, equity holders should suffer big loses, while debt holders are subordinated to any state lending (which is itself at a penal rate). After all, part of the bargain of banking is being able to convince the authorities that you should be allowed to continue; if you can't, then the game is over for the original owners and they should suffer, at very least, massive dilution. If this point is clear, at least moral hazard is confined to the treatment of debt holders[480].

It is worth noting here that the state has sometimes not struck a superb bargain for taxpayers when engaged in partial nationalisation. A comparison of the TARP recapitalisation of Goldman Sachs, for instance, with a similar injection of funds by Warren Buffett shows that the latter achieved significantly better terms[481].

One model which has been used in the past is that of a *bad bank*. This is a new firm, set up and guaranteed by the government, which buys dubious, illiquid or non performing assets as part of a clean up of the financial system. If troubled banks are first nationalised, with the bad bank being used to take bad assets, allowing the state to dispose of the cleaned-up institutions, then there is not too much moral hazard in this process[482]. But if it is used to buy assets from private financial institutions, then there is a problem in establishing the price of the purchased assets. A fair price is impossible to determine given the illiquidity of the assets. And if prices are set too high, then the bad bank is providing a covert state subsidy to bank shareholders. Taxpayers should demand a more equitable approach than this.

### 11.4.4   Product Regulation and Market Intervention

Retail products clearly require regulation to protect consumers. And the state clearly needs to set standards of business and disclosure in the wholesale markets (so that for instance insider trading is both illegal and uncommon). Other than that, though, it is not clear why one product or form of trading should be preferred over another. Certainly some recent suggestions that credit derivatives should be regulated 'as insurance' rather ignore the fact that those credit risk transfer products which were regulated as insurance – financial guarantees – caused far more problems than those that were regulated as derivatives. Risks require regulation, not particular product features.

The American mortgage market needs Fannie and Freddie at the moment: without the GSEs, there would have been few loans made in late 2008 or early 2009[483]. Whether the government should support the mortgage market in the longer term is a political question rather than one of financial stability. But certainly *if* there is a long term rôle for government support of the housing market, then it should be to the benefit of homeowners not shareholders[484]. In this context it is worth noting that both the Japanese and the GSE examples suggest that ambiguity about what is and what is not government guaranteed is unhelpful for financial stability as well as costing the taxpayer billions.

### 11.4.5   Improving Information

Economists have long known that independent experts can assist when there are problems of asymmetric information. But this requires that they are trusted by all parties, and some measure of trust has been lost in the ratings agencies after the débâcle of AAA ABS ratings. Therefore ratings agency reform is needed, whereby the agencies are paid by investors rather than issuers; they do no other business than assigning ratings; their models are not made available to the structuring banks; and there is some oversight framework to keep them honest.

### 11.4.6   Transparency, Accounting and Disclosure

One of the key lessons from the Lost Decade in Japan is that transparency matters. If impaired assets are allowed to sit around on bank balance sheets for years without losses being recognised, then investors will not trust banks for years. The rigourous application of fair value can therefore make a significant contribution to financial stability. But the readers of these fair value estimates must understand what fair values are telling them: and what they cannot tell them. In particular, market prices do not necessarily indicate the value of an asset if it is carried to term. (There is nothing to stop financial institutions who believe that fair values understate the worth of their inventory from disclosing their estimates of this higher value too.)

It is therefore rather unfortunate that, in the words of Robert Denham, chairman of the Financial Accounting Foundation[485], global accounting standards currently have 'no credibility'. His view is that the IASB is being 'jerked around by the European Commission'. Certainly the watering down of fair value that the Commission forced on the IASB hardly helps to promote investor confidence in financial institutions.

## 11.5   Experimental Finance

The last section gave some general suggestions about the reform of the rules of the financial system. But the devil is in the details. Perhaps we might start out with the best intentions, but accidentally end up with a set of rules which create a different class of crisis. A country is unlikely to volunteer to submit its banking system to an untried set of rules as an experiment in stability. So how might we get some sense of the possible behaviours that a set of rules produce?

One solution is the use of *an agent-based model* of the financial system. We have to build a reasonably representative model, so it will contain agents representing banks (with their loans and deposits); traders (including leveraged players such as hedge funds); the funding markets; and central banks. The model is based on the interactions of these agents. At the beginning we set the price of securities, interest rates, the credit quality of financial institutions, and their leverage. Then some random piece of news arrives. This changes different institutions' expectations of the future value of securities. At this point, the model allows trading to happen. If there are more buyers than sellers for an asset, then the price goes up: an excess of sellers causes the price to go down.

Meanwhile banks take deposits and make loans. The central bank sets interest rates based on the state of the economy: this affects the cost of borrowing, as does the credit quality of the borrowers. Perhaps some borrowers default: certainly price changes will cause margin calls for the leveraged borrowers which, if not met, lead to a forced liquidation of their position.

After a few rounds of this there will be gains and losses. The loan books of banks will recognise earnings on an accrual basis; securities will be marked to market. This in turn changes the amount of capital institutions have.

This is where regulatory capital comes in. We can impose a set of rules and see what happens. Doubtless if these rules are very weak, then there will be asset price bubbles and busts. Banks will fail too. But stronger rules which constrain the leverage of financial institutions will create fewer failures – perhaps asset prices will be less volatile too.

When we find a set of rules that creates good behaviours and does not allow bad ones, then perhaps these rules have some features which are worth incorporating into the regime we use in the real world. Certainly models like this one provide a good laboratory for experimenting with the financial system and its behaviours[486].

## 11.6   The Financial System from 2009

As I write, in January 2009, the third phase of the Credit Crunch may be coming to an end. State interventions are becoming larger and more dramatic: perhaps this marks the start of a fourth phase. It is worth reviewing a few key indicators to see how much worse things have become since the start of 2006. Figure 11.1 summarises a few key changes[487].

| Parameter | Start of 2006 | Start of 2009 |
|---|---|---|
| S&P 500 index | 1,248 | 903 |
| Three month USD Libor | 4.54% | 1.41% |
| FED Funds target | 4.25% | 0-0.25% |
| Case-Shiller 20 city composite index | $202K | $146K |
| U.S. retail mortgage delinquency rate | 1.62% | 6.29% |
| U.S. BBB corporate bond credit spread | 73 b.p.s. | 624 b.p.s |

Figure 11.1: Some market parameters in the Crunch

These are large moves. They demonstrate that the financial system has changed massively. How should we think about this changed financial system in 2009, what should it look like, and what should we ask of it? We end with a few suggestions.

### 11.6.1 Some changed thinking

The Credit Crunch has produced a dramatic change in thinking about the financial system and financial instruments.

Some of the threads of this revisionism are:

- A realisation that prices really can go down as well as up, and that the downs are often accompanied by rampaging illiquidity;

- A better understanding of the dangers of leverage, and of too much funding liquidity risk;

- The deepening of systems thinking as applied to finance, partly stimulated by the stubborn failure of many government interventions to get credit flowing again;

- An abandonment of blind faith in free markets, and in the efficient market hypothesis. Markets are good for some things and bad for others. Sometimes, intervention in them is required.

- In particular, the state may sometimes need to stimulate demand by printing money.

- Financial modelling is useful, but it is plagued with model risk and implicit liquidity assumptions. Modelling inherently reflexive markets is difficult, and model results should therefore be treated with scepticism.

- Innovation is not necessarily good for financial stability. It can, however, reduce the cost of credit, so there is a place for controlled development.

- There is a contrast between *transactional capitalism* (where we assume no relationship between the parties to a transaction, nor responsibilities above and beyond those defined by the transaction) and *relationship capitalism* (where there is a longer term relationship and hence short term benefit may be counterveiled by longer term considerations). The right balance between these two models is not clear. However, the primacy of the former is challenged by asymmetric information and a lack of alignment of interests. Moreover, when investors discover they have bought something that does not perform as they thought it should, lawsuits typically follow[488]. There may, then, be more of a place for relationship capitalism going forward.

## 11.6.2  The Types of Financial Institutions

It should not be governments' (nor regulators') business what types of financial institutions there are. In particular it is important that there is no incentive to be a particular type of firm when taking risk, as in the current preferential status of insurance companies versus banks for credit risk. Similarly, there is little logic to banning the ODM. But it does make sense to put controls in place so that ODM firms have appropriate capital for the risks they are taking. Perhaps models other than pure U.S. style ABS have a rôle to play here, such as the Danish model (referred to above) or covered bond issuance where the issued securities are backed by both the originator *and* the assets[489].

## 11.6.3  Society and Financial Institutions

Society will not support banks in unlimited amounts and forever. It will demand a banking system that is useful: that probably includes both a reasonable supply of credit at a reasonable price[490], and a modicum of stability. There will, in other words, be a new pact between financial institutions and society whereby in exchange for allowing them to continue taking deposits and selling products, financial institutions agree to adhere to new rules.

This book has discussed the old rules and the behaviours they led to. That sheds some light on some of the choices available in determining the terms of the new pact. These decisions should not be taken by finance ministers, regulators and senior bankers alone. They are many people's business, not least because many people are paying for the mistakes made earlier. It is important, then, that there is a broad debate about what society wants from finance, and how to get it[491]. Financial stability is too important to be left to financiers and their acolytes.

---

# Notes

[431] A further account of these phenomena can be found in Markus Brunnermeier's *Deciphering the 2007-08 Liquidity and Credit Crunch* (2008).

[432] The total estimated losses on subprime lending are only $800B, or less than 4% of the value of American equity markets. Direct subprime losses alone are far too small to explain a financial crisis of the magnitude of the current Crunch.

[433] The data comes from MDA DataQuick: see their press release *Southland home sales off bottom* (19th January 2009) available at www.dqnews.com/News/California/Southern-CA/RRSCA090119.aspx.

[434]One problem is that what is in the interest of the senior tranche holder is not necessarily what is in the interest of the junior. The senior tranche is best served by minimising losses, even if that involves writing off interest: the junior tranche is best served by maximising carry. Thus we get a situation known as *tranche warfare* where the various buyers of the SPV's debt disagree about how best to treat the collateral pool.

[435]See Al Yoon's story on Reuters of 9th January 2009, *Mortgage bonds hit by cram-down legislation advance* available at uk.reuters.com/article/marketsNewsUS/idUKN092860102 0090109.

[436]The estimate is from a presentation by Lewis Ranieri at the Milken Institute conference on Financial Innovation, April 2007.

[437]This happens when teaser rates end and when negative amortisation caps are hit in an option ARM. Borrowers in these structures are likely subprime and probably cannot refinance into another loan, especially given that their LTV is probably over 100% given the recent falls in house prices.

[438]38% is emerging as the standard for affordability. See for instance the FDIC's document *Loan Modification Program for Distressed Indymac Mortgage Loans* available at www.fdic.gov/consumers/loans/modification/indymac.html. This says that 'modifications would be designed to achieve sustainable payments at a 38 percent debt-to-income ratio'.

[439]There is a case for allowing ARMs without option features provided the mortgage rate is capped for the entire term.

[440]In particular the ability both to reassign the loan to a new property provided that the LTV criteria is still met, and to cancel the loan by buying matching MBS as well as by prepaying are important here. See Allen Frankel et al.'s *The Danish mortgage market*, BIS Quarterly Review (March 2004), available at www.bis.org/publ/qtrpdf/r_qt0403h.pdf.

[441]For one among many, see *Statement on the Government's Asset protection scheme*, UK Treasury Press Release 19th January 2009, available at www.hm-treasury.gov.uk/press_07_09.htm.

[442]For the U.S. alone by the end of 2008 the list includes the TAF, TSLF, PDCF, the FDIC's Temporary Liquidity Guarantee Program, the TARP, guarantees on assets originated by Bear Stearns, Citigroup and Merrill Lynch, the Commercial Paper Funding Facility, the Term Asset-Backed Securities Loan Facility, the ABCP Money Market Mutual Fund Liquidity Facility, the GSE Secured Lending Credit Facility and Preferred Stock Purchase Agreements, and the AIG loans and capital purchases. For the FED's perspective on its activities at the end of 2008, see the Testimony of Donald Kohn before the U.S. House of Representatives Committee on Financial Services, *Troubled Asset Relief Program* on 13th January 2009. This is available at www.federalreserve.gov/newsevents/testimony/kohn20090113a.htm.

[443]There is a definite non-linearity here. Three bailouts of $400B each seem to have less of an impact than a single $1T measure.

[444]See Bradley Keoun and Josh Fineman's Bloomberg story of 17th January 2009 *Citigroup's Pandit Tries to Save the Little That's Left to Lose*.

[445]See O. Emre Ergungor's *On the Resolution of Financial Crises: The Swedish Experience* Federal Reserve Bank of Cleveland Policy Discussion Papers Number 21 (June 2007) available at papers.ssrn.com/sol3/papers.cfm?abstract_id=1023685.

[446]Both of these are key. The RFC worked because its resources were only limited by the ability of the U.S. government to borrow, and because it could wait until bank balance sheets improved before liquidating its stake. The RFC had a measure of independence, so it could

maximise ultimate returns for taxpayers without the (politically motivated) need to get its money out early. In some cases, bank rehabilitation took more than ten years: rescuing a banking system is not necessarily speedy work. See Walker Todd's *History of and Rationales for the Reconstruction Finance Corporation*, Federal Reserve Bank of Cleveland (1992), available at `www.clevelandfed.org/research/review/1992/92-q4-todd.pdf`.

[447] It was suggested that this may be due to, in David Kyneston's phrase, 'the obeisance paid by governments of the right and left' to the financial system. Certainly the failure even to debate other forms of organisation for financial institutions than shareholder-owned corporations suggests at least a failure of imagination.

[448] The phrase comes from the British Labour Party's 1945 manifesto: they in turn borrowed it from Lenin's New Economic Plan. See Daniel Yergin and Joseph Stanislaw's *Commanding Heights* (1998).

[449] See for instance Jerry Jordan's *Financial Crises and Market Regulation*, Federal Reserve Bank of Cleveland, October 1999 available at `www.clevelandfed.org/research/comment ary/1999/1001.pdf` or Richard Bookstaber's *A Demon of Our Own Design: Markets, Hedge Funds, and the Perils of Financial Innovation* (2007): the debate here goes back to Minksy and Fisher. A comprehensive database of crises is discussed in Luc Laeven and Fabian Valencia's *Systemic Banking Crises: A New Database* IMF Working Paper 08/224 available at `www.imf.org/external/pubs/ft/wp/2008/wp08224.pdf`.

[450] For more details see William Fogarty's *Formal Investigation into the Circumstances Surrounding the Downing of Iran Air 655 on 3rd July 1988*, a declassified partial version of which is available at `www.dod.mil/pubs/foi/reading_room/172.pdf`.

[451] The Aegis combat system used by the Vincennes had been designed for cold war deep water battles, not the cluttered water and airways of the Straits of Hormuz where it was deployed: see Luke Swartz's *Overwhelmed by Technology: How did user interface failures on board the USS Vincennes lead to 290 dead?*, available at `www-cs-students.stanford.edu/~lswartz/vincennes.pdf` and the references therein for more details.

[452] Testimony before the Waxman Committee Hearings, October 2008. This is available via `oversight.house.gov/story.asp?ID=2256`.

[453] For a longer discussion see the Counterparty Risk Management Policy Group's *Containing Systemic Risk: The Road to Reform* (2008) available via `www.crmpolicygroup.org/docs/CRMPG-III.pdf`.

[454] Buying AAA ABS is a kind of tail risk position: you will only suffer if the credit enhancement is not effective. But if it does not work, the consequences are likely to be severe.

[455] One way of looking at this is to consider the process of tranching. Suppose the capital requirement on a \$1B portfolio of assets is \$60M at some confidence interval. Measuring risk at this confidence interval is equivalent to saying that any tranche attached over \$60M is risk free.

[456] VAR models were originally designed to measure market risk in the early 1990s, when the key challenge was aggregating the various interest rate risks of swaps and government bonds portfolios.

[457] The figures come from JP Morgan's 2007 Annual Report.

[458] See *The financial crisis and the future of financial regulation*, speech by Adair Turner, 21st January 2009, available at `www.fsa.gov.uk/pages/Library/Communication/Speeches/2009/0121_at.shtml`.

[459] Notice that termination agreements – whereby the protection ends with a payment of its current mark to market value if counterparty credit quality declines too far – only work if the terminated protection can be replaced. Perhaps it is time to consider a form of default swap whereby the protected asset can be delivered if the counterparty falls in quality as well as if there is a credit event.

[460] See his article *Liquidity Risk Management* in the Banque de France Financial Stability Review Special Issue on Liquidity No. 11 (February 2008).

[461] A good example of such an assumption is the strategy of Delta hedging ABS.

[462] For a more extended discussion, see FSA Consultative Paper 08/22, *Strengthening liquidity standards* (December 2008).

[463] In particular this means that transfer pricing for risky assets based on Libor flat cost of funds is inappropriate.

[464] Note that ODM banks do in fact have an incentive to originate good assets due to ware-housing risk: they have to hold the assets until they can sell them. However in (yet another) implicit liquidity assumption, this risk was ignored. Therefore the least that we can ask is that banks keep some of the risk they have originated to term. At least half of the tranches from the bottom up to above the level of expected losses would be a sensible minimum.

[465] Alignment of interests is not a panacea, as Anil Kashyap et al. point out in *The Global Roots of the Current Financial Crisis and its Implications for Regulation* (2008). But it can't hurt.

[466] See Philip Hoffman et al.'s *Surviving Large Losses: Financial Crises, the Middle Class, and the Development of Financial Markets* (2007) for an account of the responses to financial crises in the past.

[467] As Jes Staley of JP Morgan has pointed out, money market funds are highly effective con-tagion devices. They have no capital, so they cannot afford any losses. This means that as soon as there is any rumour about a firm, these funds are forced to sell its debt to protect themselves. This selling causes credit spread widening, turning a rumour into a fact.

[468] In that if we have confidence in financial institutions, it is likely that that confidence is justified. But if not, not. We discussed this earlier in section 4.1.4.

[469] Even setting the short rate is a state intervention.

[470] The Basel Committee has proposed some changes to Basel 2, but these are relatively minor and do not address the framework's biggest defects. See the press release *Basel Committee on Banking Supervision announces steps to strengthen the resilience of the banking system* (2008) available via `www.bis.org/press/p080416.htm` for details.

[471] Indeed, capital requirements should rise with the size of the institution, not fall. Lots of smaller financial institutions, none of which are too big to fail, can potentially make for a more robust financial system than a few big ones.

[472] So for instance overall leverage constraints of total assets to shareholders' funds, and of total off balance sheet risk to shareholders' funds, would be useful, hard-to-evade controls.

[473] This is not to say that other countries do not have work to do too. In the UK, for in-stance, the deposit protection regime was fatally flawed, and tripartite supervision (whereby the Treasury, the FSA, and the Bank of England cooperated) did not work quickly enough. The growth of pan-European financial institutions has not been matched by the growth of an effective pan-European supervisory authority either. There could even be a case made for a single global supervisor, or at least better supervisory cooperation.

[474] Clearly more work needs to be done on the benefits of regulatory diversification versus the potential for arbitrage. The case for a single supervisor is currently not proven.

[475] The phrase 'cognitive regulatory capture', due to Willem Buiter, captures the issue here. See his paper *Lessons from the North Atlantic financial crisis* (2008) available at `www.nber.org/~wbuiter/NAcrisis.pdf`.

[476] As Claudio Borio and Haibin Zhu say in *Capital regulation, risk-taking and monetary policy: a missing link in the transmission mechanism?* (BIS Working Paper No. 268, 2008) "changes in interest rates and the characteristics of the central bank's reaction function can influence risk-taking, by impinging on perceptions of risks and risk tolerance".

[477] As Claudio Borio points out, credit to the private sector increased by 40% from 1996 to 2006. An increase of this size was a big warning sign. See his *The Financial Turmoil of 2007-?*, BIS Working Paper No. 251 (2008).

[478] Non default components of the spread are currently large because funding is expensive and there is not enough risk capital in the market. As this situation improves, spreads should come in. At that point the Central Bank risks delaying a recovery if it keeps liquidity premiums low via its collateral policy.

[479] Not least because it allows parties who would otherwise be potential takeover partners to wait until after the bankruptcy filing and then just cherry-pick the parts that they want, leaving any dubious assets behind.

[480] This is not to say that there is no moral hazard in bailing out debt holders, just that it is more acceptable than bailing out equity holders.

[481] In his Bloomberg story of the 10th January 2009, *Paulson Bank Bailout in 'Great Stress' Misses Terms Buffett Won*, available at `www.bloomberg.com/apps/news?pid=20601087&sid=aAvhtiFdLyaQ`, Mark Pittman relates that when the TARP invested $10B in Goldman – twice as much as Buffett – it received certificates worth only a quarter as much.

[482] Two points are important here. First, the Central Bank should keep clearly solvent banks liquid by lending them money, at an appropriate rate, against collateral. But a bad bank does not do this: it buys assets, freeing up capital as well as funding, and absolving the asset holder from the consequences of any further fall in price of the asset. Second, note that the sale of a cleaned-up 'good' bank does not necessarily imply an equity spin-off. Perhaps a depositor-owned mutual or co-operative institution may make a better contribution to financial stability than a shareholder-owned bank.

[483] Most new loans made in 2008 were either conforming or supported by the FHA. According to a Bloomberg story of 10th December 2008 by Dawn Kopecki *Fannie, Freddie May Waive Appraisals for Refinancings*, James Lockhart, Director of the Federal Housing Finance Agency said that Freddie and Fannie were responsible for 73% of all new mortgages in the first nine months of 2008. Fannie and Freddie can only buy loans with an LTV of 80% or less, or with private mortgage insurance. The latter is becoming difficult to get, and falling house prices mean that many loans are above 80% LTV. This means that going forward the government body which can buy loans at higher LTVs – the FHA – is likely to be more important. Lockhart said: 'You will probably see in the next quarter the Fannie and Freddie lines going down and FHA coming up'. See `www.bloomberg.com/apps/news?pid=20601087&sid=amYWOUdC2LdY`.

[484] Before it is agreed that home ownership is a societal good, we should at least debate the long term benefits of basing economic prosperity on rising asset prices rather than on making things or providing services.

[485] The Financial Accounting Foundation oversees the FASB. His comments are detailed in Glenn Kessler's Washington Post story of 27th December 2008 *Accounting Standards Wilt Under Pressure* available at `www.washingtonpost.com/wp-dyn/content/article/2008/12/26/AR2008122601715.html`.

[486] The conclusions of models like this will never be wholly accurate, not least because of reflexivity. Once regulators announce a new set of rules, firms will try to optimise their behaviour under this new constraint. The result may well be behaviour or structures that the model has not captured.

[487] S&P 500 and Libor data are for the first day of the year. The Case-Shiller data is from the last month of the previous year, and the mortgage delinquency data, which comes from the Federal Reserve's delinquency rates for loans and leases at all commercial banks, is for the last quarter of the previous year, as available at www.federalreserve.gov/releases/chargeoff/delallsa.htm. All data is rounded to a reasonable level of significance.

[488] For a good example of the issues that can arise here, see the case of *NECA-IBEW Health & Welfare Fund vs. Goldman Sachs*. The complaint in this case is available at securities.stanford.edu/1041/GS_01/20081211_f01c_.pdf.

[489] For more on covered bonds, see the European Covered Bond Council website at ecbc.hypo.org.

[490] In a Bloomberg story of January 5th 2009, *Banks' 'Catatonic Fear' Means Consumers Don't Get TARP Relief*, James Sterngold quotes Alan Blinder, a professor of economics at Princeton, as saying that despite the bailouts, the banks are 'just sitting on the capital ... since this shows Wall Street they're safer, but then this doesn't get you much improvement. If you're taking money from the public purse, we should get something in return'. The story is available at www.bloomberg.com/apps/news?pid=20601109&sid=aqLT6v88t.Jo.

[491] There are more contributions to the debate in my blog, deusexmacc.blogspot.com.

# Index